Lecture Notes
in Control and Information Sciences   263

Editors: M. Thoma · M. Morari

Springer
London
Berlin
Heidelberg
New York
Barcelona
Hong Kong
Milan
Paris
Singapore
Tokyo

Krzysztof Gałkowski

State-space Realisations of Linear 2-D Systems with Extensions to the General nD (n>2) Case

Springer

Author

Krzyztof Gałkowski, PhD, Habil., Prof

Institute of Control and Computational Engineering, Technical University of Zielona Góra, Podgórna Str. 50, 65-246, Zielona Góra, Poland

ISBN 1-85233-410-X Springer-Verlag London Berlin Heidelberg

British Library Cataloguing in Publication Data
Galkowski, Krzysztof
    State-space realisations of linear 2-D systems with
    extensions to the general nD (n [greater than or equal to] 2)
    case. - (Lecture notes in control and information sciences
    ;263)
    1.State-space methods 2.Linear systems
    I.Title
    003
    ISBN 185233410X

Library of Congress Cataloging-in-Publication Data
Galkowski, Krzysztof, 1949-
    State-space realisations of linear 2-D systems with extensions to the general nD (n[>]2)
case / Krzysztof Galkowski.
        p. cm. -- (Lecture notes in control and information sciences, ISSN 0170-8643 ; 263)
    On t.p. "[>]" appears as the "greater than or equal to" symbol.
    Includes bibliographical references and index.
    ISBN 1-85233-410-X (acid-free paper)
        1. System analysis. 2. State-space methods. 3. Linear systems. I. Title. II. Series.
    QA402.G33 2001
    003--dc21   00-050964

Typesetting: Camera ready by author
Printed and bound at the Athenæum Press Ltd., Gateshead, Tyne & Wear
69/3830-543210 Printed on acid-free paper SPIN 10777405

Preface

The field of multidimensional (nD) systems continues to grow and mature. In particular, there are many new approaches and applications areas which are rapidly emerging into the research literature. As example, the behavioural approach to basic systems theoretic questions for nD systems has already proved to a very powerful approach for solving long standing open questions in areas such as controllability and physically relevant definitions of poles and zeros and their implications for, say, the specification and (eventual) design of control schemes.

In the applications domain paper making processes and systems with repetitive dynamics have emerged as areas where adopting an nD systems approach has clear advantages over alternatives.

In the nD systems area, the problem of constructing state-space realizations is of fundamental importance (as it is, of course, for standard (or 1D) systems). The nD case is much harder in relative terms due to complications in the theory of the underlying (polynomial/polynomial matrix based) algebraic structure, such as the distinction in the nD case between factor primeness, minor primeness, and zero primeness, or the lack of a Euclidean algorithm. These problems are particularly acute when seeking the minimal realization.

This monograph gives a comprehensive treatment of the so-called Elementary Operations Approach (EOA) for the construction of a range of state-space realizations for 2D, and more generally, nD linear systems.

In Chapter 1, we give a brief overview of the main features of nD systems, placing particular emphasis on the differences with standard, or 1D, linear systems and currently open research questions. This is supported by references to the relevant literature and an outline of the main aims and contents of this monograph.

Chapter 2 gives a summary of the relevant mathematical tools and algorithms used.

Chapter 3 solves the basic problem of: given a bivariate polynomial, construct its companion matrix. This problem is also used to illustrate the basic operation of the EOA approach.

Chapter 4 then extends these results to the case of bivariate 2D transfer functions, i.e. for single-input single-output (SISO) linear systems.

In Chapter 5, the method is extended to the multiple-input multiple-output (MIMO) linear 1D systems, which is a necessary and non-trivial step before the 2D MIMO case in Chapter 6.

In Chapter 6 it is also demonstrated that the EOA approach frequently leads to (often) undesirable singular solutions and to prevent this the use of well variable transformation methods and, in particular, inversion and the bilinear transform is investigated.

Chapter 7 extends the basic results of the previous chapters are extended to include:

1. nD (n > 2) systems,
2. linear repetitive processes (a distinct class of 2D linear systems of both theoretical and applications interest), and
3. causal systems described by Laurent polynomials.

Finally, Chapter 8 gives a critical and comparative overview of the progress reported in this monograph and give directions for future research.

Where appropriate, the algorithms and procedures in this monograph have been illustrated by examples which make extensive use of symbolic computing and, in particular, Maple.

I am grateful to Prof. Marian S. Piekarski (Technical University of Wrocław, Poland), who introduced me to the subject of nD linear systems and encouraged me to work in this area. Special thanks are due to Prof. T. Kaczorek (Warsaw University of Technology, Poland), whose encouragement and advice over many years has been of critical importance to my work. Also I am deeply thankful to Prof Eric Rogers (University of Southampton, UK) and to Professor David Owens (University of Sheffield, UK) for many fruitful discussions, research collaboration on, in particular, repetitive processes and help in preparing the final manuscript. Similar thanks are also due to Prof. Nirmal K. Bose (Pennsylvania State University, USA). Finally, I would like to thank thank Prof. Józef Korbicz (Technical University of Zielona Góra) the head of the research group where I currently work for encouragement to proceed with this monograph and my colleague Dr. Dariusz Ucinski for his excellent proof-reading.

Last but not least, I would like to thank my parents, my wife Eva and children Philip and Agatha for their patience, understanding and support.

Zielona Góra, September 2000   *Krzysztof Gałkowski*

Table of Contents

Notation

R	set of real numbers
N	set of natural numbers
$\{a,b,\dots\}$	set of elements a, b, ...
$\varphi : X \rightarrow Y$	the operator from X to Y
R^n	real n-dimensional space
$R[s_1, s_2]$	the ring of bivariate polynomials in s_1 and s_2
s_i, s, z_i, z	complex variables
$\oplus$	direct sum
$\in$	element inclusion
$\subset$	set inclusion
$\sum$	sum
$\mathrm{Mat}_{m,n}(X)$	the set of all m by n matrices with elements from X
A^T	the transpose of the matrix A
A^{-1}	the inverse of the matrix A
$\mathbf{0}$	zero matrix
$\mathbf{0}_{mn}$	zero m by n matrix
I	unit matrix
I_m	unit m by m matrix
$\deg_s[a]$	degree of the polynomial a in the variable s
rank A	the rank of the matrix A
Augment	the operation of the matrix size augmentation
BlockAugment	block operation Augment
L	the left (row) elementary operation
R	the right (column) elementary operation
L	the block left (row) elementary operation
R	the block right (column) elementary operation

1. Introduction

The past two to three decades, in particular, have seen a continually growing interest in so-called two-dimensional (2D) or, more generally, multi-dimensional (nD) systems. This is clearly related to the wide variety of applications of both practical and/or theoretical interest. The key unique feature of an nD system is that the plant or process dynamics (input, output and state variables) depend on more than one indeterminate and hence information is propagated in many independent directions.

Many physical processes have a clear nD structure. Also, the nD approach is frequently used as an analysis tool to assist, or in some cases enable, the solution of a wide variety of problems. A key point is that the applications areas for nD systems theory/engineering can be found within the general areas of Circuits, Control and Signal Processing (and many others). An obvious approach to, for example, the control related analysis of nD systems, particularly at the conceptual level where some degree of similarity is often apparent, is to simply extend standard, *i.e.* 1D, techniques. This approach is, in general, incorrect since many common 1D techniques do not generalise. Also, there are many nD systems phenomena, which have no 1D systems counterparts.

Despite the diversity of applications areas, a very large volume of literature exists on basic nD systems research and is continually being added to, see e.g. Kaczorek (1992a), Gałkowski, Ed. (2000), Gałkowski, Wood, Eds. (2001). This often requires mathematical tools outside those required in standard (1D) linear systems theory. Space limitations prevent a comprehensive treatment of these techniques. Instead we give a summary of the main ones placing emphasis on where fundamental/essential differences with the 1D case arise.

At an "abstract" level nD systems theory sets out to examine the same basic questions as 1D theory, e.g. controllability, observability, causality, construction of state space models (realisation theory), stability and stabilisation, feedback control. The basic reason why generalising 1D results (generally) fails is due to strong mathematical difficulties with the analysis tools employed. Also, as already noted, there are many key issues associated with nD, systems, which have no 1D counterparts.

Consider now the 2D/nD systems case where transfer function representations are to be used as the analysis base. Then difficulties immediately arise here due to the complexity of the underlying ring structure, *i.e.* functions in two or more indeterminates where the underlying ring does not have a division algorithm. The existence of a division algorithm for Euclidean rings forms the basis for the algorithmic derivation of many canonical forms and solution techniques at the heart of 1D systems theory, e.g. the Smith form and the solution of 1D polynomial equations.

In the 1D case, coprimeness of polynomial matrices is a key analysis tool; see e.g. Kailath (1980). There are three forms of coprimeness for an nD polynomial matrix, termed factor, minor and zero respectively. These are all equivalent in the 1D case, but for the 2D case only minor and factor coprimeness are equivalent. For the nD case none of these concepts are equivalent (Youla, Gnavi (1979), Johnson, Pugh, Hayton (1992), Johnson, Rogers, Pugh, Hayton, Owens (1996)).

As a simple example in the 2D case, consider the polynomials $s_1 + 1$ and $s_2 + 1$. These two polynomials are factor coprime but have common zero at (-1, -1). Also, the investigation of the stability of 2D linear systems (and also nD linear systems) is greatly influenced by this situation. In particular, the difficulty is that the numerator polynomial open loop can directly influence stability! This key result was first reported by Goodman (1977) after a considerable volume of literature had appeared on stability tests based on 2D transfer function descriptions. As a result, it is possible for transfer functions with the same denominator but different numerators to exhibit different stability characteristics. The key point here is that such transfer functions have non-essential singularities of the second kind, which have no 1D analogues.

The *a priori* information available and the modelling objectives permit the choice of different model structures to describe 2D or nD systems. As a basic starting point, these representations can be classified according to whether or not an input/output structure is included, and latent (or auxiliary) variables are included. This general area is discussed further in Rocha (1990, 1996) and the relevant references.

As in the 1D case, state space models are a very important class of internal representations. In this context, the concept of the state of a system can be defined (obviously) as the memory of the system, *i.e.* the past and future evolutions are independent given the current state. Hence the concept of a state depends on which ordering is considered on the underlying grid. Commonly used models for systems recursive in the positive quadrant are the Roesser model (Roesser (1975)) and the Fornasini-Marchesini models (Fornasini, Marchesini (1978)) in all their forms (see e.g. Kurek (1985) for a generalisation of the original model in this class). The structure of these models (and others) has been extensively investigated. One of the most striking differences with the 1D case is the need to consider both a global, X, and a local, x, state. Basically, the global state in a diagonal line, L_k, denoted X_k, is defined as the collection of all local states along L_k. Concepts such as reachability, controllability and observability are then defined at both global and local levels and they can be quite distinct properties.

Minimality in the 1D case is completely characterised by reachability (or controllability) and observability of the state space model. This is another key area, which does not generalise from 1D to nD where only partial results are yet known. A key problem here is again related to the complicated structure of the underlying ring structure, *i.e.* the lack of simple methods for checking the existence of solutions of polynomial equations in more than one indeterminate. (As noted by Cayley, there are no determinant-based conditions similar to Kramer's rule for systems of 1D linear equations). Grobner basis theory (Buchberger (1970) for example) may help to solve this key problem. A more detailed discussion of this problem can be found in Chapter 3.

In the 1D case, the similarity transform applied to the state vector has a key theoretical and practical role. This is again not true for nD systems where, for example, the problem of obtaining all possible state space realisations of a given state space model cannot be solved in a similar manner (Gałkowski, (1988)). Also the synthesis problem, which is at the heart of 2D/nD circuit theory and applications, has not yet been completely solved in the 2D/nD case.

Continuing, causality is a key feature of classic dynamical systems (it must be present for physical realizability). In the 1D case, the general concept of "time" imposes a natural ordering into "past", "present" and "future". Again the situation is different in the nD case due, in effect, to the fact that some of the indeterminates have a spatial rather than a temporal characteristic. Hence causality is of less relevance than in the 1D case since it is only necessary to be able to recursively perform the required computations.

Next, we give a brief overview of the developments in nD systems, which are relevant to the work reported here. The first significant work on nD systems appeared in the early 1960's in the general area of circuit analysis. There are two papers, which are widely recognised as the "starting points". First, Ozaki and Kasami (1960) showed that positive real functions of two or more variables could be used in the analysis and synthesis of circuits with variable parameters. Then, in Ansell (1964) showed how the same approach could be used for networks of transmission lines (distributed parameter elements and lumped reactances). This work was of particular interest to electrical engineers studying high frequency networks (micro-waves). It is also of interest to note that significant contributions to systems synthesis using 2D positive real functions was also reported by Piekarski and Uruski (1972). Also, the use of this theory in the study of transmission lines led to the suggestion that it could also be applied to the study of systems with delays (Żak, Lee, Lu (1986)).

Following these early publications, in particular, great interest was generated in the synthesis of 2D and nD positive real functions or matrices under the usual circuits restrictions of passivity, loselessness etc. Key papers in this general area include Koga (1968) and many papers since then by, for example, Youla, Saito, Scanlan and Rhodes. It is also important to note that this general area is still open in many respects.

The advances in the general area of signal/image processing have created many applications for, in particular, 2D and 3D filters. These include seismology, tomography and visual data communications applications. Early analysis/design tools were largely based on input/output descriptions in the form of multidimensional difference equations with immediate links via the multidimensional z and Fourier transforms to multidimensional transfer functions (or transfer function matrices). Design tools for multidimensional filters is area in which much work has been done and the filters implemented - see, for example, Fettweis (1984). Also, prominent is work by Fettweis, see for example (1991, 1992), on the extension of classical analogue circuit theory techniques (e.g. the Kirchoff laws, passivity) together with wave filters to digital applications.

Another field for nD systems theory are processes modelled by partial differential and/or difference equations. Interesting applications reported to date include river pollution modelling, Fornasini (1991).

In contrast to the 1D case, it is possible to consider models with a so-called mixed structure. For example, Kaczorek (1994) have studied systems described by state space models, which are discrete in one direction and continuous in the other. Such models have obviously close links with linear repetitive processes, which have numerous practical applications, Rogers and Owens (1992). The essential unique feature of a linear repetitive process is a series of sweeps, or passes, through a set of dynamics defined over a fixed finite duration known as the pass length. On each pass, an output, termed the pass length, is produced which acts as a forcing function on, and hence contributes to, the next pass profile. Practical examples include long-wall coal cutting and metal rolling, Smyth (1992), and algorithmic examples include classes of so-called iterative learning control schemes, see e.g. Amann, Owens and Rogers (1998), and iterative solution algorithms for classes of non-linear dynamic optimal control problems based on the maximum principle Roberts (2000).

The main subject of this monograph is in the crucial problem of constructing state-space realizations of 2D/nD linear systems from input-output data - often given in the form of a 2D/nD transfer function description. Here, the basic philosophy followed arises from the following quotation by Fornasini (1991): "The unquestioned success of the estimation and regulation procedures in 1-D theory mainly relies on state space methods, that allow for efficient and explicit synthesis algorithms. Along the same lines, it is expected that the introduction of state space models that depend on two independent variables will eventually display concrete applications of rich body of 2-D theory".

The problem of constructing state-space realizations for 2D/nD linear systems is much more complicated than for 1D, linear systems and, amongst others, this is due to the complexity of the underlying multivariate polynomial ring. This general area has been the subject of much research effort over the years (leading to a number of competitive methods) but, in general, it is still an open research problem.

A second motivation for the general approach to be employed here is taken from Youla and Pickel (1984): "Some of the most impressive accomplishments in circuits and systems have been obtained by in-depth exploitation of the properties of elementary polynomial matrices. ... Algorithms for the construction of such matrices are, therefore, of both theoretical and practical importance". Based on this underlying idea, Gałkowski has developed the so-called Elementary Operations Algorithm (EOA), which is a systematic method for constructing a range of state-space realizations for 2D systems. Note also that the possibility of achieving various equivalent state-space realizations is of particular interest for 2D/nD systems since, as mentioned, unlike the 1D linear systems case, the similarity transform does not provide all possible realizations.

Currently, there is no research text, which treats this key area in depth. This monograph aims to fill this gap in the literature by giving a comprehensive treatment of the construction of a variety of realizations (mainly of the Roesser form) – initially for the 2D case and then, by extension, to the nD (n>2) case.

2. Preliminaries

2.1 State-space Models

One of the most commonly used state-space models for 2D and, more generally nD, discrete systems is the so-called Roesser model (1975). The intrinsic feature of this model is that the partial state vector is partitioned into two sub-vectors for 2D systems or into n sub vectors for nD systems. For 2D systems these fractions are called the vertical and the horizontal state sub-vectors. For the linear case the Roesser model has a form

$$\begin{bmatrix} x^h(i+1,j) \\ x^v(i,j+1) \end{bmatrix} = A \begin{bmatrix} x^h(i,j) \\ x^v(i,j) \end{bmatrix} + Bu(i,j), \tag{2.1}$$

$$y(i,j) = C \begin{bmatrix} x^h(i,j) \\ x^v(i,j) \end{bmatrix} + Du(i,j) \tag{2.2}$$

where $i, j \in N$, N denotes the set of natural numbers, $x^h(i,j) \in R^{t_1}$ and $x^v(i,j) \in R^{t_2}$ denote the so-called horizontal and a vertical partial state vectors, $y(i,j) \in R^q$ and $u(i,j) \in R^p$ denote the output and input vectors, respectively, and A, B, C and D are compatibly dimensioned matrices with real constant entries. Finally, the initial conditions, i.e. $x^h(0,j) = \hat{x}^h(j), j = 0,1,2,...$, $x^v(i,0) = \hat{x}^v(i)$, $i = 0,1,2,...$ must be defined in order for the system response to a given input sequence to be computed. The quadruple of matrices $\{A,B,C,D\}$ is often termed the state-space realization of the Roesser model structure.

The second commonly used class of models are those of the Fornasini-Marchesini (1978) type. The most basic model here has the form

$$x(i+1, j+1) = A_1 x(i+1, j) + A_2 x(i, j+1) + A_3 x(i, j) + Bu(i, j)$$

$$\tag{2.3}$$

$$+ B_1 u(i+1, j) + B_2 u(i, j+1),$$

$$y(i, j) = Cx(i, j) + Du(i, j), \tag{2.4}$$

where $x(i, j) \in R^n$ is the partial state vector, the output and input vectors are as above and the initial conditions are now $x(0, j) = \hat{x}(j), j = 0,1,2,...,$ $x(i,0) = \hat{x}(i), i = 0,1,2,....$ Note that if A_3 and B in (2.3) are both equal to zero, the model is termed first-order and if B_1 and B_2 are both zero the model is said to be of the second order.

In what follows, a singular state space model is characterised by the presence of a singular square (or even rectangular) matrix pre-multiplying the derivative (the continuous case) or shifted (the discrete case) part of the state equation. Hence, the singular version of the Roesser model for 2D discrete linear systems has the form

$$E\begin{bmatrix} x^h(i+1, j) \\ x^v(i, j+1) \end{bmatrix} = A\begin{bmatrix} x^h(i, j) \\ x^v(i, j) \end{bmatrix} + Bu(i, j),\tag{2.5}$$

and the output equation is unchanged, see (2.2) where all notations are as in the standard Roesser model of (2.1)-(2.2). The matrix E can be square singular (or even rectangular) and note that if $E = I_N$ where $N = t_1 + t_2$ then (2.5) and (2.2) describe a standard model. Similarly, the singular version of the Fornasini-Marchesini mode has the form

$$E'x(i+1, j+1) = A_1x(i+1, j) + A_2x(i, j+1) + A_3x(i, j) + Bu(i, j)$$

$$\tag{2.6}$$

$$+ B_1u(i+1, j) + B_2u(i, j+1)$$

and again the output equation of (2.4) is unchanged. In these models singularity is as per the Roesser model above and $E' \in \text{Mat}_{n,n}(R)$ where $\text{Mat}_{n,n}(R)$ denotes the ring of $n \times n$ matrices over the field of real numbers.

Note also that the Roesser and Fornasini-Marchesini models are not completely independent of each other. In particular, it is well known that introducing the substitution

$$\xi(i, j) = x(i, j+1) - A_1x(i, j)\tag{2.7}$$

into the second-order standard Fornasini-Marchesini model gives the following Roesser model form (Kung *et al* (1977))

$$\begin{bmatrix} \xi(i+1, j) \\ x(i, j+1) \end{bmatrix} = \begin{bmatrix} A_2 & A_3 + A_2A_1 \\ I_n & A_1 \end{bmatrix}\begin{bmatrix} \xi(i, j) \\ x(i, j) \end{bmatrix} + \begin{bmatrix} B \\ 0 \end{bmatrix}u(i, j),\tag{2.8}$$

where $\xi(i, j)$ and $x(i, j) \in R^n$ represent now horizontal and vertical states, respectively. In this case, it is clear that $t_1 = t_2 = n$ and the dimension of the final Roesser model has increased by a factor of two. In the singular case it is also possible to use this method. Hence, define

$$\tilde{\xi}(i, j) = E'x(i, j+1) - A_1x(i, j)\tag{2.9}$$

which yields the following singular Roesser model

$$\begin{bmatrix} I_n & -A_2 \\ 0 & E' \end{bmatrix} \begin{bmatrix} \tilde{\xi}(i+1,j) \\ x(i,j+1) \end{bmatrix} = \begin{bmatrix} 0 & A_3 \\ I_n & A_1 \end{bmatrix} \begin{bmatrix} \tilde{\xi}(i,j) \\ x(i,j) \end{bmatrix} + \begin{bmatrix} B \\ 0 \end{bmatrix} u(i,j).$$
(2.10)

Fornasini and Marchesini (1978) also showed that their first order model is transformed to the Roesser model if

$$A_1 = \begin{bmatrix} *_{t_1 t_1} & *_{t_1 t_2} \\ 0_{t_2 t_1} & 0_{t_2 t_2} \end{bmatrix}, A_2 = \begin{bmatrix} 0_{t_1 t_1} & 0_{t_1 t_2} \\ *_{t_2 t_1} & *_{t_2 t_2} \end{bmatrix}, B_1 = \begin{bmatrix} *_{t_1 p} \\ 0_{t_2 p} \end{bmatrix}, B_2 = \begin{bmatrix} 0_{t_1 p} \\ *_{t_2 p} \end{bmatrix},$$
(2.11)

where $*_{\alpha\beta}$ and $0_{\alpha\beta}$ denote an arbitrary or a zero matrix of the assigned dimensions, respectively. Now, both the models are of the same size and the limitations are quite serious. The other algebraic methods for recasting the Roesser and the Fornasini-Marchesini models are detailed in (Gałkowski, 1996b), which use particular types of similarity and rational variable transformations.

It is also possible to define so-called continuous versions of the Roesser and Fornasini Marchesini 2D discrete linear systems state-space models. These take the forms

$$\begin{bmatrix} \dfrac{\partial}{\partial t} x^h(t,\tau) \\ \dfrac{\partial}{\partial \tau} x^v(t,\tau) \end{bmatrix} = A \begin{bmatrix} x^h(t,\tau) \\ x^v(t,\tau) \end{bmatrix} + Bu(t,\tau),$$
(2.12)

$$y(t,\tau) = C \begin{bmatrix} x^h(t,\tau) \\ x^v(t,\tau) \end{bmatrix} + Du(t,\tau),$$
(2.13)

where $t, \tau \in R$, and

$$\frac{\partial}{\partial t} \frac{\partial}{\partial \tau} x(t,\tau) = A_1 \frac{\partial}{\partial t} x(t,\tau) + A_2 \frac{\partial}{\partial \tau} x(t,\tau) + A_3 x(t,\tau) + Bu(t,\tau)$$
(2.14)

$$+ B_1 \frac{\partial}{\partial t} u(t,\tau) + B_2 \frac{\partial}{\partial \tau} u(t,\tau),$$

$$y(t,\tau) = Cx(t,\tau) + Du(t,\tau),$$
(2.15)

respectively, where all remaining notations are the same as for the discrete case.

In the classic 1D case the system can be discrete or continuous, although there are the so-called hybrid systems that are compositions of discrete and continuous fractions, see e.g. De La Sen (1997). In 2D or nD systems it is natural that one independent variable belongs to N and the second to R, and hence the system is discrete along one direction and continuous along the second (see, for example, the work of Kaczorek (1992a)). For example the discrete-continuous Roesser model can be rewritten in the form

$$\begin{bmatrix} \dfrac{\partial}{\partial t} x^h(t,j) \\ x^v(t,j+1) \end{bmatrix} = A \begin{bmatrix} x^h(t,j) \\ x^v(t,j) \end{bmatrix} + Bu(t,j),$$

(2.16)

$$y(t,j) = C \begin{bmatrix} x^h(t,j) \\ x^v(t,j) \end{bmatrix} + Du(t,j),$$

(2.17)

where $t \in R$, and $i, j \in N$. Similarly, for the Fornasini-Marchesini model

$$\frac{\partial}{\partial t} x(t,j+1) = A_1 \frac{\partial}{\partial t} x(t,j) + A_2\ x(t,j+1) + A_3 x(t,j)$$

(2.18)

$$+ Bu(t,j) + B_1 \frac{\partial}{\partial t} u(t,j) + B_2 u(t,j+1),$$

$$y(t,j) = Cx(t,j) + Du(t,j).$$

(2.19)

The 2D systems state-space models discussed above generalise naturally to the nD ($n>2$) case. The resulting Roesser state-space model is

$$\dot{x}(i_1,i_2,\cdots,i_n) = Ax(i_1,i_2,\cdots,i_n) + Bu(i_1,i_2,\cdots,i_n),$$

(2.20)

$$y(i_1,i_2,\cdots,i_n) = Cx(i_1,i_2,\cdots,i_n) + Du(i_1,i_2,\cdots,i_n),$$

(2.21)

where $x(i_1,i_2,\cdots,i_n)$, $y(i_1,i_2,\cdots,i_n)$, $u(i_1,i_2,\cdots,i_n)$ denotes the state, output and input vectors, respectively, and

$$x(i_1,i_2,\cdots,i_n) = \begin{bmatrix} x_1(i_1,i_2,\cdots,i_n) \\ x_2(i_1,i_2,\cdots,i_n) \\ \vdots \\ x_n(i_1,i_2,\cdots,i_n) \end{bmatrix},$$

(2.22)

$$\dot{x}(i_1,i_2,\cdots,i_n) = \begin{bmatrix} x_1(i_1+1,i_2,\cdots,i_n) \\ x_2(i_1,i_2+1,\cdots,i_n) \\ \vdots \\ x_n(i_1,i_2,\cdots,i_n+1) \end{bmatrix},$$

(2.23)

$x_i(i_1,i_2,\ldots,i_n)\in R^{t_i}$, $i=1,2,\ldots,n$ are the state sub-vectors; $i_1,i_2,\ldots,i_n\in N$; A, B, C, and D denote the matrices of the respective sizes. The nD linear systems Fornasini-Marchesini state-space model has the form (see e.g. Kaczorek (1999))

$$x(\Im+V)=A_0 x(\Im)+\sum_{j=1}^{n}A_j x(\Im+e_j)+\sum_{j=1}^{n}A_{1,\ldots,j-1,j+1,\ldots,n}x(\Im+V-e_j)$$

(2.24)

$$+B_0 u(\Im)+\sum_{j=1}^{n}B_j u(\Im+e_j)+\sum_{j=1}^{n}B_{1,\ldots,j-1,j+1,\ldots,n}u(\Im+V-e_j)$$

$$y(\Im)=Cx(\Im)+Du(\Im) \qquad\qquad \forall i_1,i_2,\ldots,i_n\geq 0 \qquad (2.25)$$

where

$$\Im=(i_1,i_2,\ldots,i_n),\ i_j\in N,\ j=1,2,\ldots,n\,,$$

$$V=(1,1,\ldots,1)\,,\qquad e_j=(\underbrace{0\ \cdots\ 1\cdots\ 0}_{j})\,. \qquad (2.26)$$

Finally, nD Roesser and Fornasini-Marchesini singular models are defined by multiplying the left-hand sides of the state equations in (2.20) and (2.24) by appropriately dimensioned singular matrices.

2.2 Transfer Function Matrices and Polynomial Description

As in the 1D linear case, relationships between polynomial matrix theory and state space descriptions are very strong in both the 2D/nD linear case. In the 1D discrete case the link is via the 'z' transform and in the differential case it is the Laplace transform (s). In the 2D (nD) case the multivariate 'z' or 's' (or 's' and 'z') transforms can be applied to obtain the transfer function (matrix) description, which is now a multivariate rational function (matrix).

The following transfer function matrix includes the models detailed above in the previous section as special cases

$$\aleph(s_1,s_2)=D+CH(s_1,s_2)^{-1}(B_1 s_1+B_2 s_2+B)\,, \qquad (2.27)$$

where s_1 and s_2 are complex variables. To obtain the Roesser model, set $B_1=B_2=0$, and the corresponding system matrix $H(s_1,s_2)$ is defined by

$$H(s_1,s_2)=E\Lambda-A\,, \qquad (2.28)$$

where

$$\Lambda = s_1 I_{t_1} \oplus s_2 I_{t_2},$$ (2.29)

and $\oplus$ denotes the direct sum (for a standard model E is a identity matrix). In the case of the Fornasini-Marchesini model the system matrix is defined by

$$H(s_1, s_2) = E's_1 s_2 - A_1 s_1 - A_2 s_2 - A_3.$$ (2.30)

The matrix A in (2.28) is termed the companion matrix for a bivariate polynomial $a(s_1, s_2)$, which is a common multiple of the denominators of the entries in the transfer function matrix, and in the case of the Roesser model we have

$$\det(E\Lambda - A) = a(s_1, s_2).$$ (2.31)

For the Fornasini-Marchesini model

$$a(s_1, s_2) = \det(E's_1 s_2 - A_1 s_1 - A_2 s_2 - A_3),$$ (2.32)

where, as for the Roesser model, E' is an identity matrix for the standard model.

The state-space nD ($n>2$) Roesser form description of (2.20)-(2.23) is linked to the transfer function matrix description via

$$\aleph(s_1, s_2, \cdots, s_n) = D + C(s_1 I_{t_1} \oplus s_2 I_{t_2} \oplus \cdots \oplus s_n I_{t_n} - A)^{-1} B,$$ (2.33)

which clearly is a proper rational (transfer) function matrix in several complex variables

$$\aleph(s_1, s_2, \cdots, s_n) = \left[\frac{a_{ij}(s_1, s_2, \cdots, s_n)}{b_{ij}(s_1, s_2, \cdots, s_n)} \right]_{i=1,2,\cdots,q; j=1,2,\cdots p},$$ (2.34)

where $a(.)$ and $b(.)$ are real, n-variate polynomials.

Refer now to the system matrix first introduced by Rosenbrock (1970), which is central to large areas of linear systems theory. One such area for which it provides the basic solution is the equivalence of two descriptions of a system. Recent developments in this problem for standard, or 1D, systems can, for example, be found in Johnson et al (1992) and Pugh et al (1994).

The system matrix concept can also be extended to nD linear systems, i.e. the transfer functions involved are functions of n indeterminates (most commonly $n=2$). In general, however, the results are not as powerful, in terms of explaining fundamental dynamic behaviour/properties, due to the complexity of the underlying ring structure, i.e. the ring of functions in $n>2$ indeterminates. An important exception (see Johnson et al (1996)) are linear repetitive processes (Rogers and Owens (1992)), which have practical applications in, for example, the modelling of long-wall coal cutting and metal rolling operations (Smyth (1992)). Also, the counterpart in the 2D case of Rosenbrock's work in 1D has been developed by Guiver and Bose (1982), who showed that computation of the coprime polynomial matrix fraction description of a 2D MIMO system can be obtained through computations only in the ground field of coefficients of the bivariate rational transfer function matrix.

For the standard (1D) linear systems the system matrix is defined as

$$P(s) = \begin{bmatrix} A(s) & B(s) \\ -C(s) & D(s) \end{bmatrix},$$
(2.35)

where

$$A(s) \in \text{Mat}_{t,t}(R[s]), \qquad B(s) \in \text{Mat}_{t,p}(R[s]), \qquad C(s) \in \text{Mat}_{q,t}(R[s]),$$

$$D(s) \in \text{Mat}_{q,p}(R[s]),$$

where t, p, q denote the dimension of the state-, input-, and output-vector respectively, and $\text{Mat}_{\alpha,\beta}(\Xi)$ denotes the set of $\alpha \times \beta$ matrices over the ring Ξ. It is clear that certain choices for the state equations yield the following system matrix:

$$P(s) = \begin{bmatrix} sI_t - A & -B \\ -C & -D \end{bmatrix},$$
(2.36)

where

$$A \in \text{Mat}_{t,t}(R), B \in \text{Mat}_{t,p}(R), C \in \text{Mat}_{q,t}(R), D \in \text{Mat}_{q,p}(R).$$

The matrix $\begin{bmatrix} A & B \\ C & D \end{bmatrix}$ is also called the realization matrix.

It is straightforward to generalise the previous considerations to the 2D and nD cases. Thus, the system matrix of (2.35) becomes now (2D):

$$P(s_1, s_2) = \begin{bmatrix} A(s_1, s_2) & B(s_1, s_2) \\ -C(s_1, s_2) & D(s_1, s_2) \end{bmatrix}$$
(2.37)

and (2.36)

$$P(s_1, s_2) = \begin{bmatrix} s_1 I_{t_1} \oplus s_2 I_{t_2} - A & -B \\ -C & -D \end{bmatrix},$$
(2.38)

c.f. (2.27)-(2.28) for the Roesser model. Also, note that the transfer function matrices can be represented in the polynomial matrix form as

$$F(s_1, s_2) = N_L(s_1, s_2) D_R(s_1, s_2)^{-1} = D_L(s_1, s_2)^{-1} N_R(s_1, s_2),$$
(2.39)

where $N_L(s_1, s_2)$, and $D_R(s_1, s_2)$ are right co-prime and $D_L(s_1, s_2), N_R(s_1, s_2)$ are left co-prime bivariate polynomial matrices through computations in the ground field of coefficients as detailed by Guiver and Bose (1982).

2.3 The Gałkowski Approach –the SISO Case

Gałkowski (1994) has proposed the basis of an alternative approach to the state-space realization problem for the special case of single-input / single output (SISO) 2D linear systems. The key result is that determining a state-space realization of a given SISO 2D rational transfer function can be replaced by the construction of a 3D companion matrix for a trivariate polynomial. Also, this result generalises, *i.e.* an n-variate rational function can be replaced by a $n+1$-variate polynomial. This, in turn, opens up applications of 2D systems theory to standard (1D) SISO transfer functions.

Consider first the 1D proper rational transfer function

$$f(s) = \frac{b(s)}{a(s)}, \tag{2.40}$$

i.e. the degree of the denominator is at least equal to that of the numerator polynomial, and define the following polynomial:

$$a_f(s,z) = za(s) - b(s), \tag{2.41}$$

where z is an additional complex variable. Then it is easy to see (Gałkowski (1981)) that the polynomial $a_f(s,z)$ has the companion matrix $H = \begin{bmatrix} A & B \\ C & D \end{bmatrix}$, *i.e.*

$$a_f(s,z) = \det(sI_t \oplus z - H), \tag{2.42}$$

such that

$$f(s) = D + C(sI_t - A)^{-1}B. \tag{2.43}$$

Hence, the matrix $(sI_t \oplus z - H)$ of (2.42), which is the characteristic matrix for the polynomial $a_f(s,z)$, is termed the system matrix according to (2.37), (2.38).

Consider now the proper rational 2D transfer function

$$f(s_1,s_2) = \frac{b(s_1,s_2)}{a(s_1,s_2)}, \tag{2.44}$$

and define the polynomial

$$a_f(s_1,s_2,z) = za(s_1,s_2) - b(s_1,s_2), \tag{2.45}$$

where z is an additional complex variable. Note that the notion proper transfer function means for the 2D, and generally for the nD case, that the function is proper in each variable, *i.e.* the degree in each variable of the transfer function denominator polynomial in this variable is at least equal to that of the numerator polynomial in the same variable. Then it is easy to see (Gałkowski (1981)) that the

polynomial $a_f(s_1, s_2, z)$ has the companion matrix $H = \begin{bmatrix} A & B \\ C & D \end{bmatrix}$, i.e. such that

$$a_f(s_1, s_2, z) = \det\left(E\left(s_1 I_{t_1} \oplus s_2 I_{t_2}\right) \oplus z - H\right), \tag{2.46}$$

where

$$f(s_1, s_2) = \frac{b(s_1, s_2)}{a(s_1, s_2)} = D + C\left(E\left(s_1 I_{t_1} \oplus s_2 I_{t_2}\right) - A\right)^{-1} B. \tag{2.47}$$

The transfer function (2.47) is clearly related to the Roesser state space model in both the standard and singular cases. Hence, the matrix $E\left(s_1 I_{t_1} \oplus s_2 I_{t_2}\right) \oplus z - H$, which is clearly the characteristic matrix for the polynomial $a_f(s_1, s_2, z)$, can be termed the system matrix. This leads immediately to the following approach to the realization problem. First use the transfer function $f(s_1, s_2)$ to construct the polynomial $a_f(s_1, s_2, z)$ and then solve (2.46) with the entries in the matrix H as indeterminates. Also it follows immediately that the resulting polynomial equation can be rewritten as a multivariate algebraic / polynomial equation set where on each side the coefficients associated with the same products $s_1^k s_2^l z$ are equal. This formulation is, despite the difficulty of solving the underlying equation set, turns out to be very useful in many applications. For example, it is the basis for the construction of the Elementary Operations Algorithm, which is the main subject of this monograph. This result also generalizes, i.e. an n-variate transfer function can be replaced by an $n + 1$-variate polynomial. Extension to the multiple-input, multiple-output (MIMO) systems is also possible as detailed later in this monograph.

2.4 Singularity

The singularity problems, which arise, are intrinsically related to the so-called principality, first introduced by Pontriagin (1955) and next used in the 2D/nD systems area by Bose (1982) and Lewis (1992) and defined as follows. In particular, suppose that $a(s_1, s_2)$ satisfies

$$\deg[a(s_1, s_2)] = \deg_{s_1}[a(s_1, s_2)] + \deg_{s_2}[a(s_1, s_2)], \tag{2.48}$$

where the degree of the bivariate polynomial $a(s_1, s_2)$ is defined as the degree in s of $a(s, s)$ and $\deg_{s_i}[a(s_1, s_2)]$ denotes the degree in s_i, $i = 1, 2$. If the companion matrix and the state space realization of both the forms given above have characteristic polynomials, which are principal, they are also termed principal. This concept is extremely important since it means that the Laurent expansion at infinity of the inverse of the corresponding transfer function matrix can be written as

$$H^{-1}(s_1, s_2) = s_1^{-1} s_2^{-1} \sum_{i=-\mu_1}^{\infty} \sum_{j=-\mu_2}^{\infty} \varphi_{ij} s_1^{-i} s_2^{-j}, \qquad (2.49)$$

where the lower limits on the summations are finite. This fact, in turn, permits the introduction of a transition matrix and a controllability (observability) matrix in the same way as in the standard case (Kaczorek (1992b)). An interesting case of the so-called repetitive processes described by the singular 2D state-space models is reported e.g. in Gałkowski, Rogers and Owens (1998) and Gałkowski (2000b)

Next we give some further relevant points concerning singularity in the context of this monograph. Consider first the possible cases of singularity which can arise in 1D systems. In particular, consider the transfer function $f(s) = \dfrac{s-1}{2s-1}$, which satisfies the physical realizability condition that the degree of the denominator polynomial is at least equal to that of the numerator polynomial. Clearly, this is nothing more than the standard algebraic notion of proper rational functions (matrices). Note also that the associated polynomial $a_f(s, z)$ *i.e.* $a_f = z(2s-1) - s + 1$ is principal. In such a case, there exists a variety of standard (nonsingular), as well as singular realizations, e.g.

$$E = \begin{bmatrix} 0 & 0 \\ 0 & 1 \end{bmatrix}, \; A = \begin{bmatrix} 1 & 1 \\ -1 & 0 \end{bmatrix}, \; B = \begin{bmatrix} -1 \\ 2 \end{bmatrix}, \; C = \begin{bmatrix} 1 & 0 \end{bmatrix}.$$

Consider now the improper rational function $f(s) = \dfrac{s^2 + 1}{s+1}$, which does not clearly satisfy the physical realizability condition. The associated polynomial a_f is now given by $a_f = z(s+1) - s^2 - 1$ and is not principal and hence no standard realization exists for it. It is easy to confirm, however, that its singular realization exists, which is given by the matrices

$$E = \begin{bmatrix} 0 & 0 & 0 \\ 0 & 1 & 0 \\ 0 & 0 & 1 \end{bmatrix}, \; A = \begin{bmatrix} 0 & -1 & -1 \\ -1 & 0 & 0 \\ 0 & 1 & 0 \end{bmatrix}, \; B = \begin{bmatrix} 0 \\ 0 \\ -1 \end{bmatrix}, \; C = \begin{bmatrix} -1 & 1 & 0 \end{bmatrix}.$$

This example demonstrates that transfer functions, which do not satisfy the physical realizability condition (improper), can only be realized in the singular form. Moreover the following theorem immediately follows.

Theorem 2.1 *The transfer function f is proper if and only if its associated polynomial a_f is principal.*

Proof It is easy to show that an improper transfer function produces a non-principal associated polynomial and, conversely, a non-principal polynomial can be associated only to an improper transfer function. ∎

Return now to 2D systems and consider, for example, the following singular characteristic matrix:

$$a(s_1,s_2) = \det\left(\begin{bmatrix} e_{11} & e_{12} & e_{13} & 0 \\ k_1 e_{11} & k_1 e_{12} & k_1 e_{13} & 0 \\ k_2 e_{11} & k_2 e_{12} & k_2 e_{13} & 0 \\ 0 & 0 & 0 & 1 \end{bmatrix}\begin{bmatrix} s_1 & & & 0 \\ & s_1 & & \\ & & s_2 & \\ 0 & & & s_2 \end{bmatrix} - H\right). \tag{2.50}$$

It is straightforward to show that only non-principal polynomials of the form

$$a(s_1,s_2) = a_{11}s_1 s_2 + a_{20}s_2{}^2 + a_{12}s_1 + a_{21}s_2 + a_{22} \tag{2.51}$$

can be obtained. Hence it can be concluded that if the characteristic polynomial of a 2D system is non-principal then only a singular state space realization can be found.

In the 2D/nD case, the notion of properness and the related problem of the principality of the associated polynomial matrix is more complicated since now Theorem 2.1 is not valid. Consider, for example, the transfer function

$$f(s_1,s_2) = \frac{s_1 + 1}{s_1^2 + s_2^2 + 1},$$

which is clearly proper but its denominator is not principal, and hence, the associated polynomial a_f is not principal too. Next, consider the transfer function

$$f(s_1,s_2) = \frac{s_1 + 1}{s_2 + 1}.$$

If $s_1 \equiv s_2 \equiv s$ the 1D case is recovered and it is clear that such a transfer function is proper and the associated polynomial is principal. If this is not the case, then the numerator is of degree one in the variable s_1 and of degree zero in the variable s_2 and the converse is true for the denominator polynomial. Hence, in the case of s_1, the degree of the numerator is greater than that of the denominator and the 2D physical realizability condition does not hold, i.e. the transfer function is not proper. Also, the associated polynomial a_f is given by

$$a_f = z(s_2 + 1) - s_1 - 1,$$

which is clearly not principal. This suggests that only a singular realization is possible. Note now that there is also the choice for the matrix $E(s_1 I_{t_1} \oplus s_2 I_{t_2})$, which can be now $E(s_1 I_2 \oplus s_2)$ or $E(s_2 \oplus s_1 \oplus s_2)$ since otherwise as in the 1D case redundancy may occur either in variable s_1 or in the second.

A general condition for the existence of a standard state-space realization of a 2D linear system is given by the following well-known result (see e.g. Bose (1982), Kaczorek (1988)):

Theorem 2.2 *A necessary and sufficient condition for the existence of a standard state-space realization of a 2D SISO system is that the transfer function is proper, and the associated polynomial a_f is principal.*

Proof Extending (2.46) we have that

$$a_f(s_1, s_2, z) = \det\left(E'\left(s_1 I_{t_1} \oplus s_2 I_{t_2} \oplus z\right) - H\right) \tag{2.52}$$

and it follows immediately that the standard realization can be obtained if, and only if, the associated polynomial $a_f(s_1, s_2, z)$ is principal. ∎

This theorem also generalizes to the MIMO 2D case but the principality of the multi-variate polynomial matrix related to the given transfer function matrix is more complicated and is not uniquely determined. However, it is always related to the requirement that every transfer function matrix entry is proper and a denominator of every transfer function matrix entry is principal. As shown in Chapter 5, there are many various multi-variate polynomial matrix representations for a given transfer function matrix which differ in the number of additional variables and the structure of common denominators. Clearly, various representations lead to different notions of principality. One of the possibilities is as follows:

Definition 2.1 *A MIMO nD linear system described by its transfer function matrix is principal if all the main diagonal entries in the polynomial representation of the transfer function matrix are principal and the degrees in all variables of the non-main diagonal entries do not exceed the degrees of the corresponding main diagonal entry both row and column wise.*

The other possibility is that: the polynomials belonging to the first row (column) of the polynomial matrix representation are principal and the degrees in all the variables of the remaining entries do not exceed the degrees of the corresponding first row (column) entry column (row) wise.

Consider now the following transfer function matrix:

$$F(s_1, s_2) = \begin{bmatrix} \dfrac{1}{s_1 s_2 + 1} & \dfrac{s_2}{s_1 + 1} \\ \dfrac{s_1}{s_2 + 1} & \dfrac{1}{s_1 s_2 - 1} \end{bmatrix}.$$

It is easy to see that this 2D transfer function matrix is not principal (in the sense of Definition 2.1) and hence it may be that only singular state space realizations exist for it.

3. The Elementary Operation Algorithm for Polynomial Matrices

3.1 Basics

The use of elementary operations has a long history in systems theory. An important early example is the work by Aplevich (1974).

To present the basic idea of the Elementary Operation Algorithm, consider first the simplest situation, *i.e.* the problem of how to obtain a range of companion matrices for a given bivariate polynomial. It is a straightforward exercise to show that the concept of a companion matrix is intrinsically related to the state-space realizations (see (2.28)). In particular, the companion matrix of the transfer function denominator is the matrix A which, in effect, describes the state transition in the Roesser model.

In order to generate an arbitrary 2-D companion matrix for any bivariate polynomial we will use 2-D elementary operations over the ring of bivariate polynomials $R[s_1, s_2]$, which, as commonly known, leaves a matrix determinant unchanged with respect to some nonzero constant. In this part and in all further considerations we omit this constant, as it does not play any role for our purposes. Note also that the ring of bivariate polynomials is not Euclidean and there is no direct dependence between equivalence of the bivariate monomial matrices Λ - H and Λ - G and the similarity of the matrices H and G as in the 1-D case.

Here, the notation for elementary operations plays an essential role. One possibility was indicated by Kaczorek and presented *e.g.* in (1992a). With a possible aim of employing symbolic algebraic tools for solving such problems, the Maple V notation is also given.

L(i×c) or mulrow(A,i,c)	multiplication of the ith row by the scalar $c \in R$,
R(i×c) or mulcol(A,i,c)	multiplication of the ith column by $c \in R$,
L(i+j×b(.)) or Addrow(A,j,i, b(.))	addition to the ith row of the jth row multiplied by the polynomial b(.),

R($I+j\times b(.)$) or	addition to the ith column of the jth
addcol($A,j,i, b(.)$)	column multiplied by the polynomial $b(.)$,
L(i,j) or swaprow(A,i,j)	interchange the ith and jth rows,
R(i,j) or swapcol(A,i,j)	interchange the ith and jth columns,

We will use both the notations. The first, *i.e.* Kaczorek's one is used in general discussions and in examples prepared manually to highlight some important points. The Maple notation is used in more complicated examples prepared using this package. In what follows, the augmenting operator

$$\text{Augment}: \text{Mat}_{n,n}(R[s_1,s_2]) \rightarrow \text{Mat}_{n+1,n+1}(R[s_1,s_2]), \tag{3.1}$$

is defined as

$$\text{Augment}(\Omega(s_1,s_2)) = \begin{bmatrix} 1 & 0 \\ 0 & \Omega(s_1,s_2) \end{bmatrix}. \tag{3.2}$$

Now, we can present the EOA for obtaining the companion matrix for a given 2-D polynomial. Let us assume that the polynomial $a(s_1,s_2)$ is given as

$$a(s_1,s_2) = \det \Psi(s_1,s_2) = \det \begin{bmatrix} a_{11}(s_1,s_2) & a_{12}(s_1,s_2) \\ a_{21}(s_1,s_2) & a_{22}(s_1,s_2) \end{bmatrix}, \tag{3.3}$$

where $a_{ij}(s_1,s_2)$, $i,j=1,2$ are bivariate polynomials. Our purpose is to construct an appropriate chain of left and right elementary operations that yield the characteristic matrix $E\Lambda$-H from the matrix

$$\Xi(s_1,s_2) = \begin{bmatrix} I_{N-2} & 0 \\ 0 & \Psi(s_1,s_2) \end{bmatrix}. \tag{3.4}$$

We would rather obtain a standard characteristic matrix Λ-H but this may be impossible. Also, we do not have to be able to predict the final dimension N of a characteristic matrix for a given polynomial $a(s_1,s_2)$ (mainly in the singular case). This is why it is necessary to augment a matrix dimension by applying subsequently the operator 'Augment'.

Consider

$$B_1(s_1,s_2) = \Psi(s_1,s_2), \tag{3.5}$$

where the matrix $\Psi(s_1,s_2)$ is defined in (3.3), and first augment the matrix size by using the operator 'Augment', *i.e.*

$$B_1'(s_1,s_2) = \text{Augment}(B_1(s_1,s_2)) \tag{3.6}$$

Next, apply to $B_1'(s_1,s_2)$ the chain of elementary operations:

$$L(2+1\times x_{21}(s_1,s_2)),\ L(3+1\times x_{31}(s_1,s_2)),$$
$$R(2+1\times x_{12}(s_1,s_2)),\ R(3+1\times x_{13}(s_1,s_2)), \tag{3.7}$$

where $x_{ij}(s_1,s_2),\ i,j=1,2$ are polynomials in s_1 and s_2. Each elementary operation is applied to the matrix resulting from the application of the previous one. Hence, the matrix

$$B_2(s_1,s_2)=\left[\begin{array}{c|cc} 1 & x_{12}(s_1,s_2) & x_{13}(s_1,s_2) \\ \hline x_{21}(s_1,s_2) & & \\ x_{31}(s_1,s_2) & & D(s_1,s_2) \end{array}\right], \tag{3.8}$$

where

$$D(s_1,s_2)=\Psi(s_1,s_2)+X(s_1,s_2),$$
$$X(s_1,s_2)=\left[\begin{array}{c} x_{21}(s_1,s_2) \\ x_{31}(s_1,s_2)\end{array}\right]\left[x_{12}(s_1,s_2)\quad x_{13}(s_1,s_2)\right], \tag{3.9}$$

results. The aim now is to choose polynomials $x_{ij}(s_1,s_2)$ such that some entries of the polynomial matrix $D(s_1,s_2)$ have:

• lower maximum degrees than $X(s_1,s_2)$ in one or both variables, and/or

• separated variables, provided that the maximum degrees of the remaining entries of $D(s_1,s_2)$ do not increase.

To study this problem further, introduce the following notation for an arbitrary bivariate polynomial of degree v_1 in s_1 and v_2 in s_2, respectively,

$$x_{ij}(s_1,s_2)=\sum_{i_1=0}^{v_1}\sum_{i_2=0}^{v_2}x_{i_1 i_2}s_1^{v_1-i_1}s_2^{v_2-i_2},\ i,j=1,2. \tag{3.10}$$

Note that an arbitrary polynomial $a_{ij}(s_1,s_2)$ can be written as

$$x(s_1,s_2)=b(s_1)c(s_2)+d(s_1,s_2), \tag{3.11}$$

where

$$b(s_1)=\left(\sum_{i_1=0}^{v_1}b_{i_1}s_1^{v_1-i_1}\right),\ c(s_2)=\left(\sum_{i_2=0}^{v_2}c_{i_2}s_2^{v_2-i_2}\right), \tag{3.12}$$

$$d(s_1,s_2)=\sum_{i_1=0}^{v_1}\sum_{i_2=0}^{v_2}d_{i_1 i_2}s_1^{v_1-i_1}s_2^{v_2-i_2} \tag{3.13}$$

and

$$d_{i_1 i_2} = x_{i_1 i_2} - b_{i_1} c_{i_2} .$$ (3.14)

It is easy to see that the degree of the polynomial $d(s_1,s_2)$ in both the variables decreases by one if

$$c_0 = 1; \quad b_{i_1} = x_{i_1 0}, \quad i_1 = 0,1,...,\nu_1; \quad c_{i_2} = \frac{x_{0 i_2}}{x_{00}}, \quad i_2 = 1,2,...,\nu_2;$$ (3.15)

Also, we can decrease the degree in only one polynomial variable, for example

$$b(s_1)= b_0 s_1^{\nu_1}, \quad c(s_2)= \left(\sum_{i_2=0}^{\nu_2} x_{0 i_2} s_2^{\nu_2 - i_2} \right),$$ (3.16)

where it is easy to see that the polynomial $d(s_1,s_2)$ has the degree in s_1 equal $\nu_1 - 1$. In what follows, (3.15) and (3.16) enable us to choose suitable polynomials $x_{ij}(s_1,s_2)$ in (3.8).

Note now that dealing with the general form of the initial matrix $B_1(s_1,s_2)$ given by (3.3) implies that we are not able to choose appropriate polynomials $x_{ij}(s_1,s_2)$ in (3.8) to decrease the degrees of all the entries of $D(s_1,s_2)$. We can handle simultaneously only each column or row and hence one of the polynomials $x_{ij}(s_1,s_2)$ has to be zero.

Example 3.1 Consider the polynomial

$$a(s_1,s_2) = s_1^4 s_2^3 - s_1^3 s_2^3 + s_1^3 s_2^2 - s_1^2 s_2^2 - s_1 s_2 + s_1^2 + s_2^2 ,$$

which can be represented as the determinant of the following polynomial matrix

$$B_1(s_1,s_2) = \begin{bmatrix} s_1^2 + s_2^2 + 1 & -s_1^3 s_2^2 + s_1^2 s_2^2 + 1 \\ s_1 s_2 + 1 & 1 \end{bmatrix}.$$

Here we are not able to lower the degrees of all three non-unitary entries unless the degree of the last (unitary) entry increases but this operation is possible for the first row or the first column. Hence, the second-stage matrix $B_2(s_1,s_2)$ given by (3.8) is

$$\begin{bmatrix} 1 & -s_1 & 0 \\ s_1 & s_2^2 +1 & -s_1^3 s_2^2 + s_1^2 s_2^2 +1 \\ s_2 & 1 & 1 \end{bmatrix} \text{ or } \begin{bmatrix} 1 & -1 & s_1^3 - s_1^2 \\ s_1^2 & s_1^2 +1 & 1 \\ 0 & s_1 s_2 +1 & 1 \end{bmatrix}.$$

There exist however initial matrices $B_1(s_1,s_2)$ where both columns (rows) can be treated simultaneously. Consider

$$\Psi(s_1,s_2) = \begin{bmatrix} \{s_1^{t_1}\} & \{s_1^{t_1-k}s_2^{l}\} \\ \{s_1^{k}s_2^{t_2-l}\} & \{s_2^{t_2}\} \end{bmatrix}$$

(3.17)

or

$$\Psi(s_1,s_2) = \begin{bmatrix} \{s_1^{k}s_2^{l}\} & \{s_1^{t_1-k}s_2^{l}\} \\ \{s_1^{k}s_2^{t_2-l}\} & \{s_1^{t_1-k}s_2^{t_2-l}\} \end{bmatrix},$$

(3.18)

where $\{s_1^{k}s_2^{l}\}$ denotes a polynomial with prescribed maximum degrees in both variables, *i.e.*

$$\{s_1^{k},s_2^{l}\} = \sum_{i_1=0}^{k}\sum_{i_2=0}^{l} a_{i_1 i_2} s_1^{k-i_1} s_2^{l-i_2}$$

(3.19)

and $a_{00} \neq 0$. These matrices play an essential role in the investigations to follow.

In the first case, we can obtain the second-stage matrix $B_2(s_1,s_2)$ as

$$\begin{bmatrix} 1 & b_{21}(s_1) & c_{12}(s_2) \\ -b_{12}(s_1) & \alpha_{11}(s_1) & d_{12}(s_1,s_2) \\ -c_{21}(s_2) & d_{21}(s_1,s_2) & \alpha_{22}(s_2) \end{bmatrix},$$

(3.20)

where

$$\alpha_{11}(s_1) = a_{11}(s_1) - b_{12}(s_1)b_{21}(s_1), \alpha_{22}(s_2) = a_{22}(s_2) - c_{12}(s_2)c_{21}(s_2),$$ (3.21)

and $d_{12}(s_1,s_2)$, $d_{21}(s_1,s_2)$ are defined as in (3.11) - (3.16). Hence, the degrees of these polynomials in both variables each decrease by one and the maximum degrees of the polynomials $\alpha_{ii}(.)$, i=1,2, are not greater than the maximal degrees of the polynomials $a_{ii}(.)$.

Example 3.2 Consider the polynomial

$$a(s_1,s_2) = s_1^3 s_2^3 - 3s_1^3 s_2^2 + 3s_1^3 s_2 + 5s_1^3 - 5s_1^2 s_2^2 + s_1^2 s_2 - 3s_1^2 + 2s_1 s_2^3 - 4s_1 s_2^2 + 4s_1 s_2$$

$$-9s_1 + 4s_2^3 - 4s_2^2 + 9s_2 + 5,$$

which can be rewritten as the determinant of the following polynomial matrix of the form of (3.17):

$$B_1(s_1,s_2) = \begin{bmatrix} 3s_1^3 + s_1^2 - 2s_1 + 1 & 2s_1^2 s_2 + s_1^2 - s_1 s_2 + 2s_1 - 3s_2 - 1 \\ s_1 s_2^2 + s_2^2 - 2s_1 s_2 + s_1 - s_2 + 3 & s_2^3 - 2s_2^2 + s_2 + 2 \end{bmatrix}.$$

Now, perform the operations

Augment($B_1(s_1,s_2)$), $\mathrm{L}(2+1\times(-2s_1^2+s_1+3))$, $\mathrm{L}(3+1\times(-s_2^2+2s_2-1))$,

$\mathrm{R}(2+1\times(s_1+1))$, $\mathrm{R}(3+1\times(s_2+0.5))$

to obtain

$$B_2(s_1,s_2)=\begin{bmatrix} 1 & s_1+1 & s_2+0.5 \\ -2s_1^2+s_1+3 & s_1^3+2s_1+4 & 2.5s_1+0.5 \\ -s_2^2+2s_2-1 & s_2+2 & -0.5s_2^2+s_2+1.5 \end{bmatrix}.$$

Here the degrees of the polynomials $d_{12}(s_1,s_2)$ and $d_{21}(s_1,s_2)$ have been decreased by one in both variables and for $\alpha_{22}(s_1,s_2)$ in one variable. Moreover, the degree of $\alpha_{11}(s_1,s_2)$ does not increase (see (3.11)-(3.14)).

In the second case, when employing the representation (3.18), we can proceed in a similar manner.

Example 3.3 Consider the polynomial

$$a(s_1,s_2)=s_1^5s_2^5+2s_1^5s_2^3+s_1^5s_2^2+s_1^5+s_1^4s_2^2+2s_1^3s_2^5-2s_1^3s_2^3-s_1^3-s_1^2s_2^5$$

$$+s_1^2s_2^2+s_1^2-s_1-s_2^5+s_2^3+s_2^2$$

which can again be presented as the determinant of the following polynomial matrix:

$$B_1(s_1,s_2)=\begin{bmatrix} s_1^2s_2^2+s_1^2+s_2^2-1 & s_1^2s_2^3-s_1+1 \\ s_1^3s_2^2-1 & 2s_1^3s_2^3+s_1^3-s_2^3+1 \end{bmatrix}.$$

Now perform the following operations:

step 1: Augment($B_1(s_1,s_2)$), $\mathrm{L}(2+1\times s_1^2)$, $\mathrm{L}(3+1\times(s_1^3))$, $\mathrm{R}(2+1\times(-s_2^2))$,

$\mathrm{R}(3+1\times(-2s_2^3))$ $\rightarrow B_2(s_1,s_2)$,

to obtain

$$\begin{bmatrix} 1 & -s_2^2 & -2s_2^3 \\ s_1^2 & s_1^2+s_2^2-1 & -s_1+1 \\ s_1^3 & -1 & s_1^3s_2^2+s_1^3-s_2^3+1 \end{bmatrix}.$$

Here the degrees of the polynomials $d_{12}(s_1,s_2)$ and $d_{21}(s_1,s_2)$ have been decreased in both variables and for $\alpha_{11}(s_1,s_2)$ the total degree has been decreased. Moreover, the degree of $\alpha_{22}(s_1,s_2)$ does not increase.

To proceed further, we would like to continue this procedure in such a way that the maximal degrees of the polynomial matrix entries continue decreasing. Unfortunately, for matrices of dimension greater than 2 it is not possible to deal with multiple columns and/or rows. Instead, at each step, we treat only one column or row.

Suppose, therefore, the algorithm has been used a number of times to produce the $k \times k$, $k > 2$, 2-D polynomial matrix

$$B_k(s_1, s_2) = \left[b_{ij}^k(s_1, s_2) \right]_{i,j=1,2,\ldots,k+1}. \tag{3.22}$$

Also, introduce Augment($B_k(s_1, s_2)$) and perform the elementary operations

$$R((l+1)+1 \times x_{1,l+1}(s_1,s_2)), L(2+1 \times x_{21}(s_1,s_2)), \\ L(3+1 \times x_{31}(s_1,s_2)),\ldots,L((k+2)+1 \times x_{k+2,1}(s_1,s_2)) \tag{3.23}$$

or

$$L((l+1)+1 \times x_{l+1,1}(s_1,s_2)), R(2+1 \times x_{12}(s_1,s_2)), \\ R(3+1 \times x_{13}(s_1,s_2)),\ldots,R((k+2)+1 \times x_{1,k+2}(s_1,s_2)). \tag{3.24}$$

In the first case we obtain

$$B_{k+1}(s_1,s_2) = \left[b_{ij}^{k+1}(s_1,s_2) \right]_{i,j=1,2,\ldots,k+2}$$

$$= \begin{bmatrix} 1 & 0 & \cdots & x_{1,l+1} & \cdots & 0 \\ x_{21} & b_{11}^k & \cdots & b_{1l}^k + x_{21}x_{1,l+1} & \cdots & b_{1,k+1}^k \\ \cdots & \cdots & \cdots & \cdots & \cdots & \cdots \\ x_{k+2,1} & b_{k+1,1}^k & \cdots & b_{k+1,l}^k + x_{k+2,1}x_{1,l+1} & \cdots & b_{k+1,k+1}^k \end{bmatrix} \tag{3.25}$$

and in the second

$$B_{k+1}(s_1,s_2) = \left[b_{ij}^{k+1}(s_1,s_2) \right]_{i,j=1,2,\ldots,k+2}$$

$$= \begin{bmatrix} 1 & x_{12} & \cdots & x_{1,k+2} \\ 0 & b_{11}^k & \cdots & b_{1,k+1}^k \\ \cdots & \cdots & \cdots & \cdots \\ x_{l+1,1} & b_{l1}^k + x_{12}x_{l+1,1} & \cdots & b_{l,k+1}^k + x_{1,k+2}x_{l+1,1} \\ \cdots & \cdots & \cdots & \cdots \\ 0 & b_{k+1,1}^k & \cdots & b_{k+1,k+1}^k \end{bmatrix} \tag{3.26}$$

$l=1,2,\ldots,k+1$. The polynomials x_{ij} are derived in the much the same way as for the first stage in order to decrease the degree of the respective polynomials in s_1 or s_2.

The aim, in what follows, is to obtain

$$B_\mu(s_1, s_2) = E\Lambda - H. \tag{3.27}$$

and hence the first steps of the procedure should be devoted to separate variables in all the entries of the polynomial matrix. Next, still decreasing the maximal degrees of polynomials, which is obviously connected with an augmentation of the matrix size, we can obtain a monomial matrix such that each element is a monomial in one variable (s_1 or s_2). Also, we would like to obtain a monomial matrix in which only one strict monomial entry is contained in each column and row, and the remaining matrix elements are numbers. This should be achieved by using certain non-augmenting elementary operations. Then, we could derive the final standard form $\Lambda - H$ by employing appropriate row and column permutations.

It is also possible to perform certain non-augmenting elementary operations between arbitrary sequential steps of the algorithm, which could simplify the procedure. Note, however, that it is very difficult to present the general step of the recursion in a precise way. This is due to the fact that at virtually every step different options exist leading to different final system matrices. Hence, some clarifying examples are given before the general case is again considered.

Example 3.4 Consider again the polynomial of Example 3.2, where the first step of the algorithm has been stated. Continuing the procedure gives now:

  step   2:   Augment($B_2(s_1, s_2)$),   L($4+1\times(s_2-2)$),   R($2+1\times(s_2)$), R($3+1\times(-1)$), R($4+1\times(0.5s_2)$) $\rightarrow B_3(s_1, s_2)$,

  step   3:   Augment($B_3(s_1, s_2)$),   L($4+1\times(s_1)$),   R($3+1\times(2s_1-1)$), R($4+1\times(-s_1^2-2)$), R($5+1\times(-2.5)$) $\rightarrow B_4(s_1, s_2)$,

  step 4: Augment($B_4(s_1, s_2)$), L($2+1\times(s_1)$), R($4+1\times(-2)$), R($5+1\times(s_1)$) $\rightarrow$ $B_5(s_1, s_2)$.

Finally, apply L($4+1\times(-1)$) and R($5+1\times(s_1)$) without augmentation of the matrix size. Then, by performing the appropriate row and column permutations, we obtain the final characteristic matrix of the standard type

$$\Lambda - H_1 = \begin{bmatrix} s_1 I_3 - H_{11} & -H_{12} \\ -H_{21} & s_2 I_3 - H_{22} \end{bmatrix} = \begin{bmatrix} s_1 & 1 & -2 & 0 & -1 & -2 \\ 0 & s_1 & 4 & 0 & 3 & -1 \\ 1 & 0 & s_1 & 0 & -2 & 1 \\ 0 & 0 & 4 & s_2-2 & -1 & 2 \\ 0 & 0 & -1 & 1 & s_2 & 0 \\ -1 & 0 & 1 & 0 & 3 & s_2-1 \end{bmatrix}.$$

Note here that the resulting matrix satisfies rank H_{12} = rank H_{21} = 2.

The important point here is that the EOA can be used in various ways and hence it is possible to obtain various system matrices for the same polynomial. The following examples illustrate this fact.

Example 3.5 Let the polynomial and the initial step polynomial matrix be the same as in Example 3.2 and change only some subsequent steps of the algorithm as follows:

step 1: Augment($B_1(s_1,s_2)$), L($2+1\times(-2s_1^2+s_1+3)$),

L($3+1\times(-s_2^2+2s_2-1)$), R($2+1\times(s_1+1)$), R($3+1\times(s_2)$) $\rightarrow B_2(s_1,s_2)$,

step 2: Augment($B_2(s_1,s_2)$), L($3+1\times(2s_1-1)$), R($2+1\times(s_1)$),

R($3+1\times(-0.5s_1^2-0.25s_1-1.125)$), R($4+1\times(-0.5s_1-1.25)$) $\rightarrow B_3(s_1,s_2)$,

step 3: Augment($B_3(s_1,s_2)$), L($2+1\times(s_1+1)$), R($3+1\times(-1)$),

R($4+1\times(0.5s_1-0.25)$), R($5+1\times(0.5)$) and L($3+2\times(-2)$)$\rightarrow B_4(s_1,s_2)$,

step 4: Augment($B_4(s_1,s_2)$), L($6+1\times(s_2+1)$), R($4+1\times(s_2-3)$),

R($5+1\times(-1)$) $\rightarrow B_5(s_1,s_2)$.

Finally, apply R(3×0.5), R(5×2). By performing appropriate row and column permutations, we obtain the final characteristic matrix

$$
\Lambda - H_2 = \begin{bmatrix}
s_1-0.5 & 10.25 & 0 & 0.25 & 3 & 0 \\
0 & s_1-0.5 & 1 & 0.5 & -1 & 0 \\
0.5 & -2.75 & s_1+1 & -0.75 & -1 & 0 \\
0 & 3 & -2 & s_2-1 & 3 & 0 \\
0 & -2 & 0 & 0 & s_2-3 & 1 \\
0 & 2 & 0 & 2 & -4 & s_2+1
\end{bmatrix}.
$$

Note that the resulting matrix satisfies again that rank H_{12} = rank H_{21} = 2.

The EOA formulated above can also be applied when the initial 2-D polynomial matrix $B_1(s_1,s_2)$ employed in the first step changes. Examples 3.4 and 3.5 have referred to the initial form of (3.17). Alternative forms are those given by (3.18) and the first step for this has been shown in Example 3.3. This example is now finished to explain the procedure used.

Example 3.6 Apply the following operations:

step 2: Augment($B_2(s_1,s_2)$), L($2+1\times1$), L$\left(4+1\times\left(-s_1^3\right)\right)$, R($4+1\times s_2^3$).

Now perform the following elementary operations without augmentation of the matrix size:

$$L(3+2\times1), \quad L(4+1\times1), \quad R(2+1\times1), \quad R(4+1\times1), \quad R(3+2\times(-1)) \rightarrow$$
$$B_3(s_1,s_2),$$

and the following standard ones with augmentations

step 3: Augment($B_3(s_1,s_2)$), $L(2+1\times(-s_2))$, $R(5+1\times s_2^2) \rightarrow B_4(s_1,s_2)$,

step 4: Augment($B_4(s_1,s_2)$), $L(5+1\times(-s_1))$, $R(4+1\times s_1)$, $R(6+1\times(-1)) \rightarrow$
$B_5(s_1,s_2)$,

step 5: Augment($B_5(s_1,s_2)$), $L(3+1\times(-s_2))$, $R(7+1\times s_2)$, $R(1\times(-1))$
$R(3\times(-1)) \rightarrow B_6(s_1,s_2)$,

step 6: Augment($B_6(s_1,s_2)$), $L(6+1\times s_2)$, $R(7+1\times s_2) \rightarrow B_7(s_1,s_2)$,

step 7: Augment($B_7(s_1,s_2)$), $L(9+1\times(-s_1^2))$, $R(6+1\times s_1) \rightarrow B_8(s_1,s_2)$,

step 8: Augment($B_8(s_1,s_2)$), $L(10+1\times s_1)$, $R(2+1\times s_2) \rightarrow B_9(s_1,s_2)$.

Finally, on applying appropriate row and column permutations, the following characteristic matrix $\Lambda - H$ is obtained:

$$\begin{bmatrix} s_1 & 0 & -1 & 0 & 1 & -2 & 2 & 0 & 0 & 0 \\ 1 & s_1 & 0 & 0 & 0 & 0 & 0 & 0 & 0 & 0 \\ 0 & 1 & s_1 & 0 & 0 & 0 & 0 & 0 & 0 & 0 \\ 0 & 0 & -1 & s_1 & 2 & -3 & 2 & 0 & 0 & 0 \\ 0 & 0 & 0 & -1 & s_1 & 0 & -1 & 0 & 0 & 0 \\ 0 & 0 & 0 & 0 & 0 & s_2 & 0 & 0 & 0 & 1 \\ 0 & 0 & 0 & 0 & 0 & 0 & s_2 & -1 & 0 & 0 \\ 0 & 0 & 0 & 0 & 0 & 0 & 0 & s_2 & -1 & 0 \\ 0 & 0 & -1 & 0 & 1 & -1 & 1 & 0 & s_2 & 0 \\ 0 & 0 & -1 & 0 & 2 & -2 & 1 & 0 & 0 & s_2 \end{bmatrix}.$$

Note that the resulting matrix satisfies again rank H_{12} = rank H_{21} = 2.

The initial stage representation of a polynomial can also be stated as the 1×1 matrix with this polynomial being its entry. Then

$$B_1(s_1,s_2) = \Psi(s_1,s_2) = \begin{bmatrix} 1 & 0 \\ 0 & a(s_1,s_2) \end{bmatrix}, \tag{3.28}$$

where

$$a_{11}(s_1,s_2) = 1,\, a_{12}(s_1,s_2) = 0,\, a_{21}(s_1,s_2) = 0,\, a_{22}(s_1,s_2) = a(s_1,s_2). \qquad (3.29)$$

In such a case we can always obtain a standard characteristic matrix Λ-H if and only if the considered polynomial $a(s_1,s_2)$ is monic/principal, *i.e.* its leading coefficient is unity.

Example 3.7 Consider again the polynomial from Example 3.2. and apply the following operation chain:

step 1: $\mathrm{L}(2+1\times(-s_1^3 - 2s_1 - 4))$, $\mathrm{R}(2+1\times(s_2^3 - 3s_2^2 + 3s_2 + 5)) \rightarrow B_2(s_1,s_2)$,

step 2: $\mathrm{Augment}((B_2(s_1,s_2))$, $\mathrm{L}(3+1\times(s_1))$, $\mathrm{L}(3+1\times(s_1))$, $\mathrm{R}(2+1\times(s_1^2 + 2))$, $\mathrm{R}(3+1\times(5s_1\,s_2^2 - s_1 s_2 + 3s_1 - 2s_2^2 + 2s_2 + 19)) \rightarrow B_3(s_1,s_2)$,

step 3: $\mathrm{Augment}((B_3(s_1,s_2))$, $\mathrm{L}(2+1\times(s_1))$, $\mathrm{R}(3+1\times(-s_1))$,

$$\mathrm{R}(4+1\times(-5s_2^2 + s_2 - 3)) \rightarrow B_4(s_1,s_2) = \begin{bmatrix} 1 & 0 & -s_1 & -5s_2^2 + s_2 - 3 \\ s_1 & 1 & 2 & -2s_2^2 + 2s_2 + 19 \\ 0 & 0 & 1 & s_2^3 - 3s_2^2 + 3s_2 + 5 \\ 0 & s_1 & -4 & 8s_2^2 - 3s_2 - 15 \end{bmatrix}.$$

At this stage the variables have been separated.

step 4: $\mathrm{Augment}((B_4(s_1,s_2))$, $\mathrm{R}(2+1\times(5s_2 - 1))$, $\mathrm{R}(3+1\times(2s_2 - 2))$,

$\mathrm{R}(4+1\times(-s_2^2 + 3s_2 - 3))$, $\mathrm{R}(5+1\times(8s_2 + 3))$, $\mathrm{L}(5+1\times(s_2)) \rightarrow B_5(s_1,s_2)$,

step 5: $\mathrm{Augment}((B_5(s_1,s_2))$, $\mathrm{L}(3+1\times(-5))$, $\mathrm{L}(4+1\times(-2))$, $\mathrm{L}(5+1\times(s_2 - 3))$, $\mathrm{L}(6+1\times8)$, $\mathrm{R}(2+1\times(s_2)) \rightarrow$

$$B_6(s_1,s_2) = \begin{bmatrix} 1 & s_2 & 0 & 0 & 0 & 0 \\ 0 & 1 & 0 & 0 & 0 & s_2 \\ -5 & -1 & 1 & 0 & -s_1 & -3 \\ -2 & -2 & s_1 & 1 & 2 & 19 \\ s_2 - 3 & -3 & 0 & 0 & 1 & 5 \\ 8 & 3 & 0 & s_1 & -4 & -15 \end{bmatrix}.$$

We have now obtained the monomial matrix with monomial elements only in one column and row. Finally, using $\mathrm{L}(3\times(-1))$ and performing appropriate permutations of columns and rows, we obtain the standard characteristic matrix of the form:

$$\Lambda\text{-}H_3 = \begin{bmatrix} s_1 I_3 - H_{11} & -H_{12} \\ -H_{21} & s_2 I_3 - H_{22} \end{bmatrix} = \begin{bmatrix} s_1 & 2 & 1 & -2 & -2 & 19 \\ -1 & s_1 & 0 & 1 & 5 & 3 \\ 0 & -4 & s_1 & 3 & 8 & -15 \\ 0 & 0 & 0 & s_2 & 1 & 0 \\ 0 & 1 & 0 & -3 & s_2-3 & 5 \\ 0 & 0 & 0 & 1 & 0 & s_2 \end{bmatrix}.$$

Some intermediate results have been shown here in order to highlight the procedure. Note again that $\text{rank} H_{21} = 1$

3.2 Analysis of the Elementary Operation Algorithm

In this section, the analysis based on the elementary operations algorithm is interpreted in terms of existing theories and its basic features are illustrated.

3.2.1 The EOA and Polynomial Matrix Description

First, let us present the following theorem.

Theorem 3.1 *Matrices* $B_{k_1}(s_1,s_2)$ *and* $B_{k_2}(s_1,s_2)$ *resulting from two different steps of the EOA have the same determinant, and hence they represent the same bivariate polynomial (with respect to some constant).*

Proof Simply note that applying elementary operations over the bivariate polynomial ring, and operation Augment preserve the matrix determinant (with respect to some constant). ∎

In what follows, we characterize the EOA in terms of a multivariate polynomial matrix description (PMD), which confirms that the method is correct, see Johnson *et al* (1992), Kaczorek (1990, 1992a), Pugh, Johnson (1992), Pugh *et al* (1994).

Definition 3.1 *Let* $J_i(s_1,s_2) \in \text{Mat}_{m_i,n_i}(R[s_1,s_2]), i=1,2$. *If an equation of the form*

$$K_1(s_1,s_2)J_2(s_1,s_2) = J_1(s_1,s_2)K_2(s_1,s_2) \tag{3.30}$$

exists, where $K_1(s_1,s_2) \in \text{Mat}_{m_1,m_2}(R[s_1,s_2])$, $K_2(s_1,s_2) \in \text{Mat}_{n_1,n_2}(R[s_1,s_2])$, *and* $K_1(s_1,s_2), J_1(s_1,s_2)$ *are zero left coprime and* $K_2(s_1,s_2), J_2(s_1,s_2)$ *are zero right coprime, then* $J_1(s_1,s_2)$ *and* $J_2(s_1,s_2)$ *are said to be zero equivalent.*

Remark: Zero left (right) coprimeness of the pair $K_1(s_1,s_2), J_1(s_1,s_2)$ ($K_2(s_1,s_2), J_2(s_1,s_2)$) means that there is no $(s_1^*, s_2^*) \in C^2$, which is a zero of all $m_1 \times m_1$ ($n_2 \times n_2$) minors of the matrix $[K_1(s_1,s_2) \, J_1(s_1,s_2)]$ ($\begin{bmatrix} K_2(s_1,s_2) \\ J_2(s_1,s_2) \end{bmatrix}$).

Theorem 3.2 *Square matrices* $B_{k_1}(s_1,s_2)$ *and* $B_{k_2}(s_1,s_2)$ *of the dimension* m_1 *and* $m_2 = m_1 + \partial$, *respectively, and resulting from two different steps of the EOA are zero equivalent.*

Proof It is easy to see from the algorithm that there exist 2-D unimodular matrices $P(s_1,s_2), Q(s_1,s_2) \in \text{Mat}_{m_2,m_2}(R[s_1,s_2])$. such that

$$\begin{bmatrix} I_\partial & 0 \\ 0 & B_{k_1}(s_1,s_2) \end{bmatrix} P(s_1,s_2) = Q(s_1,s_2)B_{k_2}(s_1,s_2), \tag{3.31}$$

where

$$B_{k_i}(s_1,s_2) \in \text{Mat}_{m_i,m_i}(R[s_1,s_2]), \ i=1,2.$$

Let

$$P(s_1,s_2) = \begin{bmatrix} P_1(s_1,s_2) \\ T_2(s_1,s_2) \end{bmatrix}, Q(s_1,s_2) = \begin{bmatrix} Q_1(s_1,s_2) \\ T_1(s_1,s_2) \end{bmatrix}, \tag{3.32}$$

where

$$P_1(s_1,s_2), Q_1(s_1,s_2) \in \text{Mat}_{\partial,m_2}(R[s_1,s_2]),$$

$$T_1(s_1,s_2), T_2(s_1,s_2) \in \text{Mat}_{m_1,m_2}(R[s_1,s_2]).$$

Then the matrix equation (3.31) can be written as

$$P_1(s_1,s_2) = Q_1(s_1,s_2)B_{k_2}(s_1,s_2), \tag{3.33}$$

$$B_{k_1}(s_1,s_2)T_2(s_1,s_2) = T_1(s_1,s_2)B_{k_2}(s_1,s_2). \tag{3.34}$$

It is obvious that (3.34) states the equivalence relation between $B_{k_1}(s_1,s_2)$ and $B_{k_2}(s_1,s_2)$, and it still remains to check the zero coprimeness of the respective matrices.

  a. left zero coprimeness. Due to the unimodularity of $Q(s_1,s_2)$, and as a simple consequence of the Quillen-Suslin theorem, *c.f.* Gałkowski (1994a, 1996a), $T_1(s_1,s_2)$ is zero coprime. Hence, the matrix $\left[B_{k_1}(s_1,s_2)T_1(s_1,s_2)\right]$ is also zero coprime.

  b. right zero coprimeness. Due to the unimodularity of $P(s_1,s_2)$ the matrix $\begin{bmatrix} P(s_1,s_2) \\ B_{k_2}(s_1,s_2) \end{bmatrix}$ is zero coprime. Hence, there exist the polynomial matrices $U(s_1,s_2)$ and $V(s_1,s_2)$ such that the following Bezout identity holds

$$U(s_1,s_2)P(s_1,s_2) + V(s_1,s_2)B_{k_2}(s_1,s_2) = I. \tag{3.35}$$

Note that we can partition the matrix $U(s_1,s_2)$ as

$$U(s_1,s_2) = \begin{bmatrix} U_1(s_1,s_2) & U_2(s_1,s_2) \end{bmatrix} \tag{3.36}$$

and such that

$$U(s_1,s_2)P(s_1,s_2) = U_1(s_1,s_2)P_1(s_1,s_2) + U_2(s_1,s_2)T_2(s_1,s_2). \tag{3.37}$$

Substituting (3.37) into (3.36) and noticing (3.33), we obtain

$$U_2(s_1,s_2)T_2(s_1,s_2) + \left[V(s_1,s_2) + U_1(s_1,s_2)Q_1(s_1,s_2)\right]B_{k_2}(s_1,s_2) = I. \tag{3.38}$$

This is just the Bezout identity for the matrices $T_2(s_1,s_2)$ and $B_{k_2}(s_1,s_2)$ and, hence, they are zero right coprime.   ■

3.2.2 The EOA and Similarity

The construction of a range of state-space realizations for a given 2-D transfer function is very important since it can lead to the best solution, termed the structure optimization. In particular, it can lead to a structure characterized by a minimal number of specified elements, such as capacitors, transformers or gyrators in classical applications or delay elements and multipliers in digital applications. In the 1-D case, it is only necessary to employ similarity and attention can also be limited to some subset of the possible transformations in order to preserve specified key characteristics. For example, orthogonal transformations can be used to preserve loselessness, passivity and symmetry. All these problems are relevant for the 2-D (and n-D, n>2) cases and hence it is natural to consider the use of similarity transformations. Previous work by Kung *et al* (1977) treated this problem but considered only block diagonal nonsingular matrices of appropriate dimensions. It is related to the standard requirements for the 1D similarity, *i.e.*

$$T^{-1}\Lambda T = \Lambda,$$
(3.39)

which is obvious since in this case $\Lambda = sI_n$. Even in the 2D case, the situation is much more complicated. Let the nonsingular matrix T be written in the block form as

$$T = \begin{bmatrix} T_{11} & T_{12} \\ T_{21} & T_{22} \end{bmatrix}$$
(3.40)

and its inverse as

$$T^{-1} = \begin{bmatrix} S_{11} & S_{12} \\ S_{21} & S_{22} \end{bmatrix}.$$
(3.41)

Then, it is straightforward to show that

$$T^{-1}\Lambda T := T^{-1} \begin{bmatrix} s_1 I_n & 0 \\ 0 & s_2 I_m \end{bmatrix} T = \begin{bmatrix} S_{11} & S_{12} \\ S_{21} & S_{22} \end{bmatrix} \begin{bmatrix} s_1 I_n & 0 \\ 0 & s_2 I_m \end{bmatrix} T$$
(3.42)

$$= \begin{bmatrix} s_1 S_{11} & s_2 S_{12} \\ s_1 S_{21} & s_2 S_{22} \end{bmatrix} T = \begin{bmatrix} s_1 S_{11} T_{11} + s_2 S_{12} T_{21} & s_1 S_{11} T_{12} + s_2 S_{12} T_{22} \\ s_1 S_{21} T_{11} + s_2 S_{22} T_{21} & s_1 S_{21} T_{12} + s_2 S_{22} T_{22} \end{bmatrix}.$$

The above suggests that the similarity matrix T can only be block diagonal with blocks of appropriate dimensions and that non-singularity of this matrix is not sufficient. In actual fact, however, this is not true since (see Gałkowski (1994a, 1996a)) given an nD companion matrix H, there can exist a non-block diagonal (full) nonsingular matrix T such that

$$\det(\Lambda - H) = \det(\Lambda - T^{-1}HT),$$
(3.43)

although (3.39) does not hold, *i.e.*

$$T^{-1}\Lambda T \neq \Lambda.$$
(3.44)

Note, however, that there are square non-singular matrices of corresponding dimensions which do not satisfy (3.43) and cannot be employed as nD similarity matrices - the first example of non-block diagonal 2D similarity matrices was given in Gałkowski (1988). In this work, it is shown that transformations, which link both the image and transpose of the matrix under consideration may be treated as similarity matrices (in the current context) and hence (3.43) and (3.44) hold simultaneously since

$$\det(\Lambda - H^T) = \det(\Lambda^T - H^T) = \det(\Lambda - H)^T = \det(\Lambda - H).$$
(3.45)

Therefore, consider the matrix pair

$$A = \begin{bmatrix} 1 & 2 & 3 \\ 4 & 5 & 6 \\ 7 & 8 & 9 \end{bmatrix}, \quad B = \begin{bmatrix} 1 & 4 & 7 \\ 2 & 5 & 8 \\ 3 & 6 & 9 \end{bmatrix}.$$

Clearly these matrices satisfy $B^T = A$ and the matrix T satisfying $A = T^{-1}BT$ is equal to

$$T = \begin{bmatrix} 1 & 0 & 0 \\ 0 & 8/9 & -2/9 \\ 0 & -2/9 & 5/9 \end{bmatrix}.$$

This result is easily established using a Maple routine, *i.e.* issimilar(A,B,`T`).

The fact now is that the matrix A can be transformed with a matrix T which is not block diagonal with compatible dimensions because transformations by block diagonal matrices T are only a particular case and do not cover all possibilities. At this stage, it is important to note that obtaining a matrix T such that (3.43) holds is a very difficult and strongly nonlinear problem. In particular, a nonsingular matrix of compatible dimensions is not sufficient. As detailed next, the EOA provides a solution to this extremely important problem.

Return to Examples 3.4, 3.5, and 3.7 that are related to the same bivariate polynomial and note that all companion matrices H_i, $i=1,2,3$ achieved in these Examples for the same polynomial $a(s_1,s_2)$ are similar in a certain 1-D sense. In particular,

$$H_2 = T^{-1}H_3T,$$

where

$$T = \begin{bmatrix} 0 & 1 & 0 & 0 & 0 & 0 \\ -0.1853 & -0.0600 & 0.5513 & 0.2014 & 0.5595 & 0.2318 \\ 0.1789 & -0.3304 & 1.4677 & 0.1731 & 0.6104 & 1.3790 \\ 0.5901 & 0.2853 & -0.5114 & -0.5139 & -1.5753 & -0.0775 \\ 0.3373 & 0.1136 & 0.1010 & -0.4233 & -0.1034 & -0.2501 \\ 0.9235 & 0.4934 & -0.4912 & -0.0632 & 0.4691 & -1.3337 \end{bmatrix}.$$

Further, the matrix T is not block diagonal with block dimensions related to Λ. Hence, it is proved that by changing the procedure in the EOA one can obtain companion matrices that are linked by non-block diagonal similarity. This can be accomplished in two ways, *i.e.* as shown previously by changing the internal elementary operations or by changing the initial matrix representation of the polynomial.

To highlight the second possibility, consider the polynomial

$$a(s_1,s_2) = s_1^2 s_2^2 + s_1^2 s_2 + 2s_1 s_2^2 + 3s_1 s_2 + s_1 + s_2 + 2,$$

which can be expressed in the form of (3.17) as

$$\Psi_1(s_1,s_2)=\begin{bmatrix} s_1^2 & 3s_1s_2+3s_1-s_2-2 \\ s_1s_2+2s_1+1 & 7s_2^2+13s_2+6 \end{bmatrix}$$

or

$$\Psi_2(s_1,s_2)=\begin{bmatrix} s_1^2 & 14s_1s_2+22s_1-2s_2-4 \\ s_1s_2+3s_1+0.5 & 15s_2^2+65s_2+66 \end{bmatrix}.$$

It is easy to use the EOA procedure to obtain, for example, the companion matrices

$$H_1 = \begin{bmatrix} 0 & 7 & -3 & -3 \\ 0 & -2 & 2 & 1 \\ 0 & -2 & 0 & 1 \\ -1 & 0 & 0 & -1 \end{bmatrix}$$

and

$$H_2 = \begin{bmatrix} 0 & 15 & -22 & -14 \\ 0 & -2 & 4 & 2 \\ 0 & -1 & 0 & 1 \\ -0.5 & -2 & 0 & -1 \end{bmatrix}$$

respectively. The subsequent steps of calculations are not given here since they do not provide any extra information. Finally, it is easy to check that

$$H_2 = T^{-1}H_1T,$$

where

$$T = \begin{bmatrix} 0.4858 & -0.2116 & -3.5588 & -2.7935 \\ 1.6627 & 9.1036 & 8.0876 & 2.1254 \\ -1.1514 & -3.0211 & -6.7054 & -2.4801 \\ 4.5655 & 18.6445 & 29.4208 & 10.1024 \end{bmatrix}.$$

3.3 The Elementary Operation Algorithm and Singularity

In general, the algorithm developed previously does not yield a standard characteristic matrix Λ-H, but a singular $E\Lambda - H$. It is sometimes possible to avoid this singularity and the following result relates to such cases.

Theorem 3.3 *For each 2-D polynomial 2×2 matrix representation $A(s_1, s_2)$ of an arbitrary bivariate polynomial, presented in (3.3), the EOA can be applied to obtain the form Δ - Π, where $\Delta = 0_k \oplus I_{\tau_1} s_1 \oplus I_{\tau_2} s_2$, $\tau_1 \geq t_1$, and $\tau_2 \geq t_2$, Π is an $M \times M$ real or complex matrix, $M = k + \tau_1 + \tau_2 \geq N$, $N = t_1 + t_2$, and 0_k is the zero square matrix of the dimension k. Hence, there exist unimodular matrices $P(s_1, s_2)$ and $Q(s_1, s_2)$ representing chains of appropriate elementary operations such that*

$$\Delta - \Pi = P(s_1, s_2) \begin{bmatrix} I_{M-2} & 0 & 0 \\ 0 & a_{11}(s_1, s_2) & a_{12}(s_1, s_2) \\ 0 & a_{21}(s_1, s_2) & a_{22}(s_1, s_2) \end{bmatrix} Q(s_1, s_2). \tag{3.46}$$

Proof Immediate on application of the elementary operations of the algorithm developed here. ∎

It worth noticing that the above theorem is also valid for an arbitrary square 2-D polynomial matrix $A(s_1, s_2)$ of the size greater than two.

Obviously, the form Δ - Π can be expressed in the usual singular form, where

$$\Delta - \Pi = E' \Lambda' - \Pi \tag{3.47}$$

and

$$E' = \begin{bmatrix} 0_k & 0 \\ 0 & I_{\tau_1 + \tau_2} \end{bmatrix}, \Lambda' = s_1 I_{k + \tau_1} \oplus s_2 I_{\tau_2}, \tag{3.48}$$

such that det $E' = 0$. Also, note that the characteristic matrix Δ - Π can be written in the block form

$$\Delta - \Pi = \begin{bmatrix} -H_{11} & -H_{12} & -H_{13} \\ -H_{21} & s_1 I_{\tau_1} - H_{22} & -H_{23} \\ -H_{31} & -H_{32} & s_2 I_{\tau_2} - H_{33} \end{bmatrix}, \tag{3.49}$$

where the submatrices of Π have compatible, relative to Δ, dimensions. Also, the singularity of the characteristic matrix Δ - Π = $E' \Lambda' - \Pi$ can be avoided when the matrix H_{11} is nonsingular. In such cases, we obtain a standard form Λ-H by using well-known properties of block determinants, *i.e.*

$$\det(\Delta - \Pi) = (-1)^k \det H_{11}$$
$$\times \det \left\{ s_1 I_{\tau_1} \oplus s_2 I_{\tau_2} - \begin{bmatrix} H_{22} & H_{23} \\ H_{32} & H_{33} \end{bmatrix} + \begin{bmatrix} H_{21} \\ H_{31} \end{bmatrix} [H_{11}]^{-1} [H_{12} \quad H_{13}] \right\}, \tag{3.50}$$

where $\tau_i = t_i, i = 1, 2$. Hence, we have obtained a standard characteristic matrix Λ - H, where

$$H = \begin{bmatrix} H_{22} & H_{23} \\ H_{32} & H_{33} \end{bmatrix} - \begin{bmatrix} H_{21} \\ H_{31} \end{bmatrix} [H_{11}]^{-1} [H_{12} \quad H_{13}] \qquad (3.51)$$

Note that we can obtain the characteristic matrix Δ - Π with a nonsingular block H_{11}, whereas the considered polynomial $a(s_1,s_2)$ is monic. Moreover, we must have used redundant operations *i.e.* it is possible to perform more operations at one step of the algorithm than we did.

If the submatrix H_{11} is singular, then we cannot obtain a standard characteristic matrix. In such a case, we deal with the strictly singular characteristic matrix $E'\Lambda'$-H. We can also mostly reduce its dimension as before. Let the submatrix H_{11} have the rank equal to k-l. Then, by premultiplying by a suitable square matrix P, we obtain

$$PH_{11} = \begin{bmatrix} \alpha_{k-l} & 0 \\ 0 & 0_l \end{bmatrix}, \qquad (3.52)$$

where α_{k-l} is some nonsingular $(k$-$l) \times (k$-$l)$ submatrix. Hence, the characteristic matrix can be written as

$$\Delta' - \Pi' = \begin{bmatrix} \alpha_{k-l} & \vdots & \multicolumn{2}{c}{\beta} \\ \hline & 0_l & \vdots & * \\ \gamma & \vdots & s_1 I_{\tau_1} - H'_{22} & -H'_{23} \\ & * & -H'_{32} & s_2 I_{\tau_2} - H'_{33} \end{bmatrix}, \qquad (3.53)$$

where * denotes an arbitrary submatrix of appropriate dimensions. Thus, using a block determinant property, we obtain

$$\det(\Delta' - \Pi') = \det \alpha_{k-l} \det \left\{ \begin{bmatrix} 0_l & \vdots & * \\ \hline * & s_1 I_{\tau_1} - H'_{22} & -H'_{23} \\ & -H'_{32} & s_2 I_{\tau_2} - H'_{33} \end{bmatrix} - \gamma \alpha^{-1} \beta \right\}. \qquad (3.54)$$

Finally, the singular characteristic matrix $E'\Lambda'$-H can be written in the same manner as in (3.49) with k replaced by $l \le k$ and

$$H_{11} = 0. \qquad (3.55)$$

Further consideration of Theorem 3.3 shows that the EOA can yield a singular characteristic matrix whose form is given by (3.54). We can obtain such a form for a monic polynomial when there also exist standard ones, as well as for non-monic polynomials when no standard $\Lambda - H$ exists. For a monic polynomial, the singularity of the final characteristic matrix relies on the initial matrix $\psi(s_1,s_2)$ employed in first step of the algorithm. Generally, this yields a singular result. This problem will also be studied in the next section devoted to the analysis of various possible initial polynomial representations.

To discuss both singular cases, *i.e.* principal and non principal, in more detail, note that on applying subsequent steps of the algorithm a matrix whose entries are univariate monomials or real numbers results. Such a matrix can contain the submatrix

$$\xi = \begin{bmatrix} \varepsilon s_2 + \alpha & \beta \\ \phi s_1 + \chi & \phi s_2 + \delta \end{bmatrix} \qquad (3.56)$$

provided that suitable row and column permutations have been performed. We can also find other blocks ξ where s_1 and s_2 are swapped, e.g.

$$\xi = \begin{bmatrix} \varepsilon s_1 + \alpha & \beta \\ \phi s_2 + \chi & \phi s_1 + \delta \end{bmatrix}. \qquad (3.57)$$

It is also possible that ε or ϕ is zero, but $\phi \neq 0$. In all these cases, we cannot obtain $\begin{bmatrix} s_1 + \alpha & \beta \\ \chi & s_2 + \delta \end{bmatrix}$ by using non-augmenting elementary operations. However, applying our approach we can augment the matrix size, e.g. we can transform (3.56) to the form

$$\begin{bmatrix} 0 & 1 & -1 & 0 \\ 1 & s_1 & 0 & 0 \\ 0 & 0 & s_2 + \alpha' & \beta' \\ -1 & 0 & \chi' & \delta' \end{bmatrix}, \qquad (3.58)$$

where

$$\alpha' = \frac{\alpha}{\varepsilon}, \quad \beta' = \frac{\beta}{\varepsilon}, \quad \chi' = \frac{\chi}{\phi}, \quad \delta' = \frac{\delta}{\phi}.$$

Now, consider the following example of a monic polynomial whose initial representation, in turn, yields singular realisations when employing the EOA:

Example 3.8 Consider the matrix

$$\Psi(s_1, s_2) = \begin{bmatrix} s_1 - 1 & s_1^2 s_2 + s_1^2 - s_1 s_2 + s_1 + s_2 - 1 \\ s_2 + 1 & 2s_1 s_2^2 - 2s_2^2 + 3s_1 s_2 - s_2 + 1 \end{bmatrix},$$

which represents the monic polynomial

$$a(s_1, s_2) = s_2^2(s_1^2 - 3s_1 + 1) + s_2(s_1^2 - 4s_1 + 1) - s_1^2.$$

It is easy to see that this is an example of the principal case. Now, apply the following algorithm:

$$B_1(s_1, s_2) = \Psi(s_1, s_2),$$

step 1: Augment($B_1(s_1, s_2)$), L($2 + 1 \times (s_1^2 - s_1 + 1)$),

L($3 + 1 \times (s_1 s_2 - 2s_2 + 1)$), R($3 + 1 \times (-s_2 - 1)$) $\rightarrow B_2(s_1, s_2)$,

step 2: Augment($B_2(s_1, s_2)$), L($3 + 1 \times (-s_1)$), L($4 + 1 \times (-2s_2 - 1)$),

R($2 + 1 \times (s_1 - 1)$), R($3 + 1 \times 1$) $\rightarrow B_3(s_1, s_2)$,

step 3: Augment($B_3(s_1, s_2)$), L($5 + 1 \times (s_2)$), R($2 + 1 \times 2$), R($4 + 1 \times 1$) $\rightarrow$
$B_4(s_1, s_2)$,

step 4: R($5 + 2 \times 2$) $\rightarrow B_5(s_1, s_2) = \begin{bmatrix} 1 & 2 & 0 & 1 & 4 \\ 0 & 1 & s_1 - 1 & 1 & 2 \\ \{0\} & 0 & 1 & 0 & \{-s_2 - 1\} \\ 0 & -s_1 & 1 & -1 & -2 \\ \{s_2\} & -1 & 2 & 0 & \{-s_1 - 2\} \end{bmatrix}$.

It is easy to see that the matrix $B_5(s_1, s_2)$ contains the block (braced entries) of the form (3.56). For avoidance of this disadvantage, perform a suitable column permutation. So, we still have to augment the matrix.

step 5: Augment($B_5(s_1, s_2)$), L($2 + 1 \times 1$), L($5 + 1 \times (-2)$), L($6 + 1 \times 1$),

R($6 + 1 \times (s_1)$) $\rightarrow B_6(s_1, s_2)$,

step 6: Augment($B_6(s_1, s_2)$), L($2 + 1 \times (s_1)$),

R($7 + 1 \times (-1)$) $\rightarrow B_7(s_1, s_2) = \begin{bmatrix} 1 & \langle 0 \rangle & 0 & 0 & 0 & \langle 0 \rangle & -1 \\ s_1 & 1 & 0 & 0 & 0 & 0 & 0 \\ 0 & \langle 0 \rangle & 1 & 2 & 0 & \langle 1 \rangle & 0 \\ 0 & 0 & 0 & 1 & s_1 - 1 & 1 & 0 \\ 0 & 0 & 0 & 0 & 1 & 0 & -s_2 - 1 \\ 0 & -2 & 0 & -s_1 & 1 & -1 & -2 \\ 0 & 1 & s_2 & -1 & 2 & 0 & 0 \end{bmatrix}$.

Applying appropriate row and column permutations to $B_7(s_1, s_2)$, multiplying some of them by "-1" and finally decreasing the matrix size as given in (3.54), where the bracketed elements have been taken as the block H_{11}, yields the representation

$$\Pi' - \Delta' = \begin{bmatrix} 0 & -1 & 0 & 0 & -1 & 0 \\ 1 & s_1 & 0 & 0 & 0 & 0 \\ 0 & 0 & s_1-1 & 1 & 0 & -1 \\ -2 & 0 & 1 & s_1-2 & 2 & 1 \\ 0 & 0 & 1 & 0 & s_2+1 & 0 \\ 1 & 0 & 2 & 1 & 0 & s_2 \end{bmatrix}.$$

The following example illustrates the non-principal polynomial case.

Example 3.9 Let us consider the polynomial

$$a(s_1,s_2) = 0s_1^3 s_2^2 - s_1^3 s_2 + s_1^2 s_2 - s_1 s_2^2 - s_1^3 + s_1 s_2 + s_1^2 + s_2^2 + s_2 - 2.$$

It is easy to check that it relates to the following matrix representation of the form (3.17):

$$a(s_1,s_2) = \det B_1(s_1,s_2) = \det \begin{bmatrix} s_1^3 + 1 & s_1 s_2 - 1 \\ s_1^2 s_2 + s_1^2 + s_2 - 1 & s_2^2 - 1 \end{bmatrix}.$$

Apply the following operations

   step 1: Augment($B_1(s_1,s_2)$), $L(2+1\times(-s_1))$, $L(3+1\times(-s_2-1))$,
   $R(2+1\times(s_1^2+1))$ $R(3+1\times s_2) \rightarrow B_2(s_1,s_2)$,

   step 2: Augment($B_2(s_1,s_2)$), $L(4+1\times s_2)$, $R(2+1\times 1)$, $R(4+1\times 1) \rightarrow$
   $B_3(s_1,s_2)$,

   step 3: Augment($B_3(s_1,s_2)$), $L(3+1\times(-s_1))$, $L(4+1\times 1)$, $R(4+1\times s_1) \rightarrow$
   $B_4(s_1,s_2)$.

Now, applying appropriate column and row permutations and multiplication by "-1", we obtain

$$B_4'(s_1,s_2) = \begin{bmatrix} s_1 & -1 & 0 & 0 & 0 \\ 1 & s_1 & -1 & 0 & s_2 \\ 1 & -1 & s_1 & 0 & -1 \\ -2 & 0 & 1 & s_2 & -1 \\ 0 & 0 & 1 & -1 & -1 \end{bmatrix} = E'\Lambda - H',$$

where

$$E' = \begin{bmatrix} 1 & 0 & 0 & 0 & 0 \\ 0 & 1 & 0 & 0 & 1 \\ 0 & 0 & 1 & 0 & 0 \\ 0 & 0 & 0 & 1 & 0 \\ 0 & 0 & 0 & 0 & 0 \end{bmatrix}, \Lambda' = s_1 I_3 \oplus s_2 I_2.$$

Now apply the following operations to $B_4'(s_1, s_2)$:

  step 4: Augment($B_4'(s_1, s_2)$), $L(3+1\times(-1))$, $R(6+1\times s_2)$

and apply suitable row permutations and multiplications by"-1" to obtain

$$\Pi' - \Delta' = \begin{bmatrix} 0 & 0 & 0 & -1 & 1 & 1 \\ 0 & s_1 & -1 & 0 & 0 & 0 \\ -1 & 1 & s_1 & -1 & 0 & 0 \\ 0 & 1 & -1 & s_1 & 0 & -1 \\ 0 & -2 & 0 & 1 & s_2 & -1 \\ 1 & 0 & 0 & 0 & 0 & s_2 \end{bmatrix}.$$

This analysis confirms the fact that there is no standard (nonsingular) companion matrix for non-principal polynomials. However, for principal polynomials we can derive both standard and singular solutions depending on the initial polynomial representation used. In Chapter 6, it will be shown how this disadvantage can be avoided by using well-defined variable transforms such as inversion or bilinear transform.

3.4 Initial Representations for the Elementary Operation Algorithm

Consider now the problem of choosing initial matrix representations for a given 2D polynomial. First, consider the diagonal initial form (3.28). Example 3.7 shows that the submatrix H_{21} has unit rank. This property is general in the sense that starting from (3.28) – then the first size augmentation is not needed - H_{21} or H_{12} must have the rank equal to one, *i.e.*

$$\text{rank}(H_{12}) = 1 \text{ and/or } \text{rank}(H_{21}) = 1. \tag{3.59}$$

This condition coincides with the fact that the 2-D characteristic matrix Λ-H, where H denotes the Sun Yuan Kung *et al* (1977) companion matrix for a bivariate polynomial a, which can be transformed to the same canonical Smith form as in the 1-D case by using elementary operations. Hence there exist unimodular matrices $P(s_1, s_2)$ and $Q(s_1, s_2)$ representing the product of elementary operations such that

$$\Lambda - H = P(s_1, s_2)\begin{bmatrix} I_{n-1} & 0 \\ 0 & a(s_1, s_2) \end{bmatrix} Q(s_1, s_2). \tag{3.60}$$

As noted previously, the initial representation (3.28) always yields the standard characteristic matrix $\Lambda - H$ but subject to the very restrictive limitations of (3.59). Fortunately, these can be weakened by employing the general 2×2 representation (3.3) in the initial step of the EOA. In fact, the standard characteristic matrix can be obtained if we use the initial representations of the polynomial $a(s_1, s_2)$ given by (3.17) or (3.18). However, employing these initial representations does not completely remove the rank limitation. In fact

$$\operatorname{rank} H_{12} \leq \min(l, t_1 - k) + 1, \operatorname{rank} H_{21} \leq \min(t_2 - l, k) + 1. \tag{3.61}$$

This is because we can treat only a few rows and columns simultaneously only in the first steps of the algorithm. In the next steps we can operate on only one row or column. This means that there are a large number of zero entries in the resulting matrix.

Augmenting again the size of the initial representation can weaken the condition (3.61). Consider, for example, an initial representation of the type

$$\Psi(s_1, s_2) = \begin{bmatrix} \left\{s_1^{t_1-1}\right\} & \left\{s_1^{t_1-1-k}s_2^l\right\} & \left\{s_1^{t_1-1-k}s_2\right\} \\ \left\{s_1^k s_2^{t_2-1-l}\right\} & \left\{s_2^{t_2-1}\right\} & \left\{s_2^{t_2-l}\right\} \\ \left\{s_1^{k+1}\right\} & \left\{s_1 s_2^l\right\} & \left\{s_1 s_2\right\} \end{bmatrix}. \tag{3.62}$$

Then, it turns out that this representation can yield standard results and the condition of (3.61) is weakened.

Example 3.10 Consider the polynomial

$$a(s_1, s_2) = s_1^3 s_2^3 + 5s_1^3 s_2^2 - 14s_1^3 s_2 - 4s_1^3 - 8s_1^2 s_2^3 + 50s_1^2 s_2^2 + 38s_1^2 s_2 + 10s_1^2 - 32s_1 s_2^3$$

$$-59s_1 s_2^2 - 34s_1 s_2 - 8s_1 + 21s_2^3 + 12s_2^2 + 10s_2 - 3$$

with the following initial representation

$$\Psi(s_1, s_2) = \begin{bmatrix} s_1^2 + s_1 + 1 & -s_1 s_2 + s_1 - s_2 - 1 & s_1 s_2 + s_1 - 2s_2 + 1 \\ 2s_1 s_2 + s_1 - 2s_2 - 2 & -s_2^2 + s_2 - 1 & 4s_2^2 + s_2 + 1 \\ s_1^2 + s_1 - 1 & -2s_1 s_2 - s_1 + s_2 - 1 & 7s_1 s_2 + 5s_1 - 5s_2 + 2 \end{bmatrix}.$$

Apply the EOA procedure:

Augment($B_1(s_1,s_2) = \Psi(s_1,s_2)$), $L(2+1\times(s_1+2))$, $L(3+1\times(2s_2+1))$,
$L(4+1\times(3s_1+4))$, $R(2+1\times(-s_1+1))$, $R(3+1\times(s_2-1))$, $R(4+1\times(-2s_2-1)) \rightarrow$
$B_2(s_1,s_2)$,

Augment($B_2(s_1,s_2)$), $L(2+1\times(-1))$, $L(3+1\times(-1))$, $L(4+1\times(-s_2))$,
$L(5+1\times(-s_1-5))$, $R(4+1\times(s_2-4))$.

Now, apply the non-augmenting operation

$R(2+1\times2) \rightarrow B_3(s_1,s_2)$,

and again

Augment($B_3(s_1,s_2)$), $L(3+1\times2)$, $L(4+1\times6)$, $L(5+1\times3)$,
$L(6+1\times(-s_1+13)$, $R(6+1\times(s_2+2))$.

Then, apply the non augmenting operations

$R(2+1\times(-1))$, $R(3+1\times1)$, $L(5+2\times(-4)) \rightarrow B_4(s_1,s_2)$.

The final multiplication of the 3-rd and 5-th rows by '-1' and applying suitable row and column permutations yields the characteristic matrix of the form

$$
\begin{bmatrix}
s_1-1 & -1 & 2 & -3 & -3 & 3 \\
3 & s_1+6 & -6 & 11 & 1 & -7 \\
3 & 7 & s_1-13 & 24 & 15 & -18 \\
0 & 1 & -1 & s_2+2 & 0 & -1 \\
0 & 2 & 0 & 0 & s_2-4 & 1 \\
1 & 4 & 3 & -6 & -14 & s_2+7
\end{bmatrix}.
$$

It is easy to see that both the submatrices H_{12} and H_{21} are now of the full rank equal 3 in this case.

The representation of (3.62) does not, however, provide the general solution since there is still the rank limitation as illustrated by the next example.

Example 3.11 Consider the polynomial of greater degrees in both the variables

$$
a(s_1,s_2) = s_1^4 s_2^4 + s_1^4 s_2^3 - 3s_1^4 s_2 - 4s_1^4 + s_1^3 s_2^4 - 8s_1^3 s_2^3 - 9s_1^3 s_2^2 - 5s_1^3 s_2
$$

$$
- s_1^3 - s_1^2 s_2^4 + s_1^2 s_2^3 - 8s_1^2 s_2^2 - 11s_1^2 s_2 + 5s_1^2 + 6s_1 s_2^4 + 18s_1 s_2^3
$$

$$
+ 13s_1 s_2^2 - 5s_1 s_2 - 11s_1 - 10s_2^4 - 7s_2^3 + 5s_2^2 + 10s_2 - 2
$$

with the similar initial representation

$$\Psi(s_1,s_2) = \begin{bmatrix} s_1^3 + s_1^2 - 2s_1 + 1 & s_1 s_2^2 + 2s_1 s_2 + s_1 + 1 & s_1 s_2 - s_1 + s_2 + 1 \\ s_1^2 s_2 + s_1 s_2 - s_2 + 1 & 2s_2^3 + s_2^2 - s_2 - 1 & s_2^2 + s_2 - 1 \\ s_1^3 - s_1^2 - s_1 + 3 & s_1 s_2^2 - s_1 + s_2^2 + 1 & 2s_1 s_2 + s_1 - s_2 + 2 \end{bmatrix}.$$

Apply the EOA procedure:

Augment($B_1(s_1,s_2) = \Psi(s_1,s_2)$),   L($2 + 1 \times s_1$),   L($3 + 1 \times s_2$),

L($4 + 1 \times (s_1 + 1)$), R($2 + 1 \times (-s_1^2 - s_1)$), R($3 + 1 \times (-s_2^2 - 2)$), R($4 + 1 \times (-s_2 + 1)$).

Now, apply the non-augmenting operations

R($2 + 1$), R($3 + 1 \times (-1)$) $\rightarrow B_2(s_1,s_2)$,

and again

Augment($B_2(s_1,s_2)$),   L($2 + 1 \times (s_1 + 1)$),   L($3 + 1 \times 1$),   L($5 + 1 \times (3s_1 + 1)$),
R($3 + 1 \times s_2$) $\rightarrow B_3(s_1,s_2)$,

Augment($B_3(s_1,s_2)$),   L($3 + 1 \times s_2$),   L($5 + 1 \times (-s_2^2 + s_2 + 2)$),
R($5 + 1 \times s_2$) $\rightarrow B_4(s_1,s_2)$,

Augment($B_4(s_1,s_2)$),   L($4 + 1 \times (-1)$),   L($6 + 1 \times s_2$),   R($2 + 1 \times (s_2 - 1)$),
R($4 + 1 \times (-1)$), R($7 + 1 \times (-2)$) $\rightarrow B_5(s_1,s_2)$,

Augment($B_5(s_1,s_2)$), L($8 + 1 \times s_1$), R($4 + 1 \times (-3)$), R($5 + 1 \times (-1)$), R($7 + 1 \times 4$),
R($8 + 1 \times (-s_2 - 2)$).

Now, apply the non-augmenting operations

L($5 + 1 \times (-1)$), L($6 + 1$), L($8 + 1 \times (-2)$) $\rightarrow B_6(s_1,s_2)$.

Final row or column multiplications by the appropriate scalars and application of suitable row and column permutations yields a characteristic matrix of the form

$$\begin{bmatrix} s_1 - 2 & 7 & 3 & 4 & -7 & 0 & 0 & -10 \\ -1 & s_1 + 4 & 3 & 1 & -3 & 1 & -1 & -7 \\ 1 & -2 & s_1 - 1 & 1 & 1 & 0 & 0 & 5 \\ 0 & 1 & 0 & s_1 & 0 & 0 & 0 & 0 \\ 1 & -3 & -1 & 0 & s_2 + 2 & 0 & 0 & 4 \\ 0 & 0 & -1 & 0 & 2 & s_2 - 1 & 1 & 0 \\ 0 & 0 & 0 & 1 & 1 & 2 & s_2 & -1 \\ 0 & 0 & 0 & 0 & 0 & 1 & 0 & s_2 \end{bmatrix}.$$

It is straightforward to see that the submatrices H_{12} and H_{21} here are not of the full rank since $\text{rank}H_{12} = \text{rank}H_{12} = 3$. Thus, the representation of (3.62) does not always yield all possible system matrices.

Consider now the following polynomial matrix:

$$\Psi(s_1,s_2) = \begin{bmatrix} \{s_1^{t_1-\alpha}\} & \{s_1^{t_1-\alpha-k}s_2^l\} & \{s_1^{t_1-\alpha-k}s_2^\beta\} \\ \{s_1^k s_2^{t_2-\beta-l}\} & \{s_2^{t_2-\beta}\} & \{s_2^{t_2-l}\} \\ \{s_1^{k+\alpha}\} & \{s_1^\alpha s_2^l\} & \{s_1^\alpha s_2^\beta\} \end{bmatrix}, \tag{3.63}$$

which is clearly a generalised version of the representation (3.62), where $\alpha = \beta = 1$. Also the more general cases when $\alpha \geq 1, \beta \geq 1$ are much more complicated and, in fact, it is relatively easy to obtain a singular result despite the fact that the polynomial under consideration is, in fact, principal. The following example highlights this key fact.

Example 3.12 Consider the bivariate polynomial matrix

$$\Psi(s_1,s_2)$$

$$= \begin{bmatrix} s_1^2 + s_1 - 1 & -2s_1 s_2 + s_1 + s_2 + 2 & s_1^2 s_2 + 2s_1^2 + s_1 + s_2 + 1 \\ 2s_1 s_2 + s_1 - s_2 - 1 & s_2^2 + 2s_2 + 2 & s_1 s_2^2 - s_2^2 - s_1 s_2 - 1 \\ s_1^2 s_2 + s_1^2 - s_1 s_2 + 1 & s_1 s_2^2 + s_1 s_2 + s_2^2 + s_2 - 1 & 2s_1^2 s_2^2 + s_1^2 s_2 + 2s_1 s_2 + s_1 - 2 \end{bmatrix}$$

Apply the following steps of the EOA:

Augment($B_1(s_1,s_2) = \Psi(s_1,s_2)$),   L($2+1\times(2s_1-1)$),   L($3+1\times(2s_2+1)$),
L($4+1\times(s_1 s_2 + s_1)$), R($2+1\times(-s_1+1)$),   R($3+1\times s_2$),
R($4+1\times(-2s_1 s_2 + s_1 - 2)$) $\rightarrow B_2(s_1,s_2)$,

Augment($B_2(s_1,s_2)$),   L($2+1\times(2s_2-1)$),   L($3+1\times(3s_1 s_2 - 4s_1 - 2s_2 + 4)$),
L($4+1\times(3s_2^2 + s_2 - 1)$), L($5+1\times(-s_1+1)$), R($3+1\times s_2$) $\rightarrow B_3(s_1,s_2)$,

where

$$B_3(s_1,s_2) = \begin{bmatrix} 1 & 0 & 0 & 0 & s_1 \\ 2s_2-1 & 1 & -s_1+1 & s_2 & -2 \\ 3s_1 s_2 - 4s_1 - 2s_2 + 4 & 2s_1 - 1 & -s_1^2 + 4s_1 - 2 & s_1 + 2 & s_2 + 3 \\ 3s_2^2 + s_2 - 1 & 2s_2 + 1 & s_2 & 3s_2^2 + 3s_2 + 2 & -\dfrac{11}{3}s_2 - \dfrac{10}{3} \\ -s_1+1 & s_1 s_2 + s_1 & 1 & s_1 s_2^2 + 2s_1 s_2 + s_2 - 1 & -2 \end{bmatrix}.$$

The last column of this matrix strongly suggests (see $(3.57) - (3.58)$) that the resulting characteristic matrix is certainly singular. Hence, the general case of the initial representation of (3.63) is not suitable for our purposes when applied alone. It can, however, be the subject of further investigations when combined with variables transforms discussed in subsequent chapters.

The problem of how to find an initial representation for the EOA, which is not associated to any rank conditions, *i.e.* can yield the final solution with arbitrary ranks of blocks H_{12} and H_{21} is still an open question and is the subject of ongoing work and will be reported in due course. Possible approaches are the methods of construction of multivariable polynomial matrices with a prescribed determinant, based on the Quillen-Suslin theorem, see e.g. Gatazzo (1991). Also, symbolic algebra methods based on Grobner bases, see e.g. Buchberger (1970) appear to be very promising. Some of the results are also shown in the next sub-section.

To complete the general problem of choosing an initial matrix representation of a bivariate polynomial, it is necessary to consider its existence and the procedure for producing it. Consequently, return to the 2×2 initial representation (3.3) for a given polynomial $a(s_1, s_2)$. Clearly, as was already mentioned the simplest is the representation of (3.28), since the given polynomial just generates it. However, this representation does not yield all possible representations. A more promising, but more complicated, situation is for representations of the type (3.17)-(3.18). This is simply a modification of the 2-D polynomial Diophantine identity

$$\alpha(s_1, s_2)x(s_1, s_2) + \beta(s_1, s_2)y(s_1, s_2) = \gamma(s_1, s_2). \tag{3.64}$$

According to (3.3), (3.64) can be rewritten as

$$a_{11}(s_1, s_2)a_{22}(s_1, s_2) + a_{12}(s_1, s_2)a_{21}(s_1, s_2) = a(s_1, s_2), \tag{3.65}$$

where it is assumed that

$$a(s_1, s_2) = \sum_{i_1=0}^{t_1} \sum_{i_2=0}^{t_2} c_{i_1 i_2} s_1^{t_1-i_1} s_2^{t_2-i_2}. \tag{3.66}$$

In the classic form of the equation the polynomials $\alpha(s_1, s_2), \beta(s_1, s_2)$ and $\gamma(s_1, s_2)$ are known and $x(s_1, s_2)$ and $y(s_1, s_2)$ are variables to be found. It is known that the zero-coprimeness of the polynomials $\alpha(s_1, s_2), \beta(s_1, s_2)$ guarantees the existence of solutions, which is equivalent to the requirement that the ideal generated by the polynomials $\alpha(s_1, s_2), \beta(s_1, s_2)$ covers the whole polynomial ring (Youla, Pickel (1984), Gatazzo (1991)). This is only a sufficient condition, since it is enough that the polynomial $\gamma(s_1, s_2)$ belongs to this ideal. In our problem, all polynomials $a_{ij}(s_1, s_2), i, j = 1,2$ in (3.3), *i.e.* $\alpha(s_1, s_2), \beta(s_1, s_2)$ and $x(s_1, s_2), y(s_1, s_2)$ can be arbitrarily chosen and only the right-hand side polynomial $a(s_1, s_2)$ (in (3.3) or (3.65)), *i.e.* the polynomial $\gamma(s_1, s_2)$ in (3.64), is known. Generally, therefore, the problem should always have solutions since it suffices to find co-prime polynomials $\alpha(s_1, s_2)$ and $\beta(s_1, s_2)$. However, we require the special form of it defined by (3.17) or (3.18), which makes the problem a little more difficult.

Consider first the representation (3.17) and note that here the polynomials $a_{11} = \left\{ s_1^{t_1} \right\}$ and $a_{21} = \left\{ s_1^k s_2^{t_2-l} \right\}$ cannot be coprime (over the complex field) since the polynomial a_{11} has infinitely many roots of the form $\left\{ s_1^*, x \right\} x \in C$ and the univariate polynomial $a_{21}^*(s_2) \triangleq a_{21}(s_1^*, s_2)$ always has some roots. Thus, we can meet some problems with this representation. However, the representation (3.18), where both the polynomials are bivariate should always exist.

The other problem consists in effective computation of a variety of solutions. Here we use a modification of the method proposed by Kaczorek (1982). Assume that the coefficients of the polynomials of one of the rows or columns of the matrix $\Psi(s_1, s_2)$ ((3.17) or (3.18)) are parameters. Then the remaining coefficients of polynomials of $\Psi(s_1, s_2)$ can be obtained as functions of these parameters from (3.65).

First, consider the initial representation of the type (3.17). For simplicity of notation, we consider the particular case of the polynomial with the degrees in both variables equal to three:

$$a_{11}(s_1) = \sum_{i_1=0}^{3} a_{i_1} s_1^{3-i_1} \ , \quad a_{12}(s_1, s_2) = \sum_{i_1=0}^{2} \sum_{i_2=0}^{1} b_{i_1 i_2} s_1^{2-i_1} s_2^{1-i_2} \ ,$$

$$a_{21}(s_1, s_2) = \sum_{i_1=0}^{1} \sum_{i_2=0}^{2} a_{i_1 i_2} s_1^{1-i_1} s_2^{2-i_2} \ , \quad a_{22}(s_2) = \sum_{i_2=0}^{3} b_{i_2} s_2^{3-i_2} \ . \tag{3.67}$$

Thus, (3.65) can now be rewritten as

$$d_j = 0, \ j = 1, 2, \dots, 16 \tag{3.68}$$

where, in Maple notation

```
> d1:=a0*b0-a00*b00-c00:

> d2:=a0*b1-a10*b00-a00*b10-c10:

> d3:=d3=a0*b2-a20*b00-a10*b10-c20:

> d4:=d4=a0*b3-a20*b10-c30:

> d5:=a1*b2-a21*b00-a11*b10-a20*b01-a10*b11-c21:

> d6:=a1*b3-a21*b10-a20*b11-c31:

> d7:=a2*b2-a21*b01-a11*b11-a20*b02-a10*b12-c22:

> d8:=a2*b3-a21*b11-a20*b12-c32:
```

```
> d9:=a3*b2-a21*b02-a11*b12-c23:

> d10:=a3*b3-a21*b12-c33:

> d11:=a1*b0-a01*b00-a00*b01-c01:

> d12:=a1*b1-a11*b00-a01*b10-a10*b01-a00*b11-c11:

> d13:=a2*b0-a01*b01-a00*b02-c02:

> d14:=a2*b1-a11*b01-a01*b11-a10*b02-a00*b12-c12:

> d15:=a3*b0-a01*b02-c03:

> d16:=a3*b1-a11*b02-a01*b12-c13:.
```
$$(3.68a)$$

Using the Maple command `solve' to solve the equation set (3.68) is not really feasible and this comment also holds for the particular polynomial of (3.66), *i.e.* for the coefficients c00 to c33. Note also that obtaining necessary and sufficient conditions for the solution of a set of nonlinear algebraic equations (here they are actually bilinear) is much more complicated than in the linear case. Firstly, there does not exist the well-known Kramer rule, which provides effective solution methods, see e.g. Lancaster (1969). When equations are not linear but multilinear, or nonlinear, such a rule does not exist in general. In has been proven by Caley [Cambridge and Dublin Mathematical Journal, vol. III, p. 116] that, given a set of n multi–linear forms with n indeterminates there exists no an equation set such that its determinant is their resultant. This problem has only particular solutions, for example:

a. Sylvester solved the problem for $n=3$ (Cambridge and Dublin Mathematical Journal, vol. VII, p. 68),

b. Versluys (see Baraniecki 1879) – for arbitrary n, but only one form is square and the rest are linear, or

c. Gundelfinger (*Schloemilch Zeitschrift fur Math. u Ph.*, vol. XVIII p. 549.) – for n arbitrary, but two forms are square and the rest are linear.

Hence, classic algebra does not provide an effective solution for this problem. However, some insights can be reached using the Grobner bases, see Buchberger (1970)).

Theorem 3.4 *(Buchberger (1970)). The equation set is solvable if, and only if, the associated Grobner basis does not contain the unit element.* ∎

In general, the use of symbolic computing to solve of this problem is a very difficult task. For example, applying the Maple subroutine for Grobner bases to the equation set (3.68) has failed to produce the solution. In such a case, the standard numerical method can be proposed.

Method 1: Introduce the square sum

$$\mathbf{D} = \sum_{i=1}^{16} d_i^2 .$$
$$(3.69)$$

Then, the solution of (3.68) can be achieved as follows:

Find a_i, a_{ij} such that $\mathbf{D}$ attains a minimum value equal to zero.

As stated previously, this problem can be easily solved numerically. A symbolic solution for arbitrary values c00 to c33 is expected to be a very complicated function of these parameters, and hence pointless. Also, some mixed methods can be used, *i.e.* partially symbolic, partially numerical.

Method 2: Introduce the following coefficients vectors:

$$c^T = \begin{bmatrix} c_{00} & c_{10} & c_{20} & c_{30} & \vdots & c_{21} & c_{31} & c_{22} & c_{32} & c_{23} & c_{33} & c_{01} & c_{11} & c_{02} & c_{12} & c_{03} & c_{13} \end{bmatrix}^T ,$$

$$b^T = \begin{bmatrix} b_0 & b_1 & b_2 & b_3 & \vdots & b_{00} & b_{10} & b_{01} & b_{11} & b_{02} & b_{12} \end{bmatrix}^T . \tag{3.70}$$

Then, we can rewrite (3.64) as linear vector equation

$$Xb = c , \tag{3.71}$$

where

$$X = \left[\begin{array}{cccc:cccccc}
a_0 & 0 & 0 & 0 & -a_{00} & 0 & 0 & 0 & 0 & 0 \\
0 & a_0 & 0 & 0 & -a_{10} & -a_{00} & 0 & 0 & 0 & 0 \\
0 & 0 & a_0 & 0 & -a_{20} & -a_{10} & 0 & 0 & 0 & 0 \\
0 & 0 & 0 & a_0 & 0 & -a_{20} & 0 & 0 & 0 & 0 \\ \hdashline
0 & 0 & a_1 & 0 & -a_{21} & -a_{11} & -a_{20} & -a_{10} & 0 & 0 \\
0 & 0 & 0 & a_1 & 0 & -a_{21} & 0 & -a_{20} & 0 & 0 \\
0 & 0 & a_2 & 0 & 0 & 0 & -a_{21} & -a_{11} & -a_{20} & -a_{10} \\
0 & 0 & 0 & a_2 & 0 & 0 & 0 & -a_{21} & 0 & -a_{20} \\
0 & 0 & a_3 & 0 & 0 & 0 & 0 & 0 & -a_{21} & -a_{11} \\
0 & 0 & 0 & a_3 & 0 & 0 & 0 & 0 & 0 & -a_{21} \\ \hdashline
a_1 & 0 & 0 & 0 & -a_{01} & 0 & -a_{00} & 0 & 0 & 0 \\
0 & a_1 & 0 & 0 & -a_{11} & -a_{01} & -a_{10} & -a_{00} & 0 & 0 \\
a_2 & 0 & 0 & 0 & 0 & 0 & -a_{01} & 0 & -a_{00} & 0 \\
0 & a_2 & 0 & 0 & 0 & 0 & -a_{11} & -a_{01} & -a_{10} & -a_{00} \\
a_3 & 0 & 0 & 0 & 0 & 0 & 0 & 0 & -a_{01} & 0 \\
0 & a_3 & 0 & 0 & 0 & 0 & 0 & 0 & -a_{11} & -a_{01}
\end{array} \right] . \tag{3.71a}$$

A necessary and sufficient condition for the solvability of (3.71) is

$$\operatorname{rank} X = \operatorname{rank}\left[X \mid c \right]. \tag{3.72}$$

Our aim now is to find values for a_i and a_{ij} such that the condition of (3.72) holds.

Then, it is easy to conclude that the matrix X is of the maximal rank, *i.e.* equal to 10, if the coefficients a_0 and a_{21} or a_{01} are non-zero.

In what follows, we apply the symbolic procedure from Maple for LU decomposition

```
> U1:=LUdecomp(X,P='P1',L='L1'):
```
(3.73)

which is also very time-consuming but provides accurate solutions. Next, derive

```
> H1:=inverse(P1&*L1):
```

```
> c1:=evalm(H1&*c):
```
(3.74)

and note that (3.71) has been transformed to the form

$$U_1 b = c_1, \tag{3.75}$$

where the last six rows of U_1 are zero. Thus, (3.75) has a solution if and only if the last six elements of the column c_1 are zero too. Hence, instead 16 bilinear equations to solve, we have only 6, but these are of higher degrees and hence more complicated, polynomial equations, which can be, however, solved by using Method 1.

In what follows, we present also methods related to the initial representation of the form of (3.18), which gives equivalent solutions. Now, the polynomials of (3.67) have the form

$$a_{11}(s_1,s_2)=\sum_{i_1=0}^{1}\sum_{i_2=0}^{1} a_{i_1 i_2} s_1^{1-i_1} s_2^{1-i_1} \ , \quad a_{12}(s_1,s_2)=\sum_{i_1=0}^{2}\sum_{i_2=0}^{1} f_{i_1 i_2} s_1^{2-i_1} s_2^{1-i_2} \ ,$$

$$a_{21}(s_1,s_2)=\sum_{i_1=0}^{1}\sum_{i_2=0}^{2} e_{i_1 i_2} s_1^{1-i_1} s_2^{2-i_2} \ , \quad a_{22}(s_1,s_2)=\sum_{i_1=0}^{2}\sum_{i_2=0}^{2} b_{i_1 i_2} s_1^{2-i_1} s_2^{2-i_2} \ . \tag{3.76}$$

Also in this case, the polynomial equation (3.65) can be rewritten in the form of the bilinear equation set of (3.68), but now d_i, $i=1,2,...,16$ can be written down as

```
>d1:=a00*b00-f00*e00*-c00:

>d2:=a01*b00+a00*b01-f00*e01-f01*e00-c01:

>d3:=a01*b01-f00*e02+a00*b02-f01*e01-c02:

>d4:=a01*b02-f01*e02-c03:

>d5:=-f00*e10-f10*e00+a10*b00+a00*b10-c10:

>d6:=a01*b10-f01*e10+a00*b11-f00*e11-f10*e01+a11*b00
+a10*b01-f11*e00-c11:

>d7:=-f01*e11+a11*b01+a01*b11-f00*e12-f11*e01-f10*e02
+a00*b12+a10*b02-c12:

>d8:=a01*b12-f01*e12-f11*e02+a11*b02-c13:
```

```
>d9:=f10*e10-f20*e00+a00*b20+a10*b10-c20:

>d10:=f21*e00-f10*e11+a00*b21+a11*b10+a01*b20+a10*b11-
f11*e10-f20*e01-c21:

>d11:=-f11*e11+a11*b11-f21*e01+a00*b22-f10*e12+a01*b21
+a10*b12-f20*e02-c22:

>d12:=a01*b22-f21*e02-f11*e12+a11*b12-c23:

>d13:=-f20*e10+a10*b20-c30:

>d14:=a10*b21-f20*e11+a11*b20-f21*e10-c31:

>d15:=a11*b21-f21*e11+a10*b22-f20*e12-c32:

>d16:=a11*b22-f21*e12-c33:                                    (3.77)
```

This equation set can be solved numerically by using Method 1, and also Method 2 can be modified to apply to this equation set. In this latter case, the matrix X is defined as

$$
\begin{bmatrix}
-e_{00} & 0 & 0 & 0 & 0 & 0 & a_{00} & 0 & 0 & 0 & 0 & 0 & 0 & 0 & 0 \\
-e_{01} & -e_{00} & 0 & 0 & 0 & 0 & a_{01} & a_{00} & 0 & 0 & 0 & 0 & 0 & 0 & 0 \\
-e_{02} & -e_{01} & 0 & 0 & 0 & 0 & 0 & a_{01} & a_{00} & 0 & 0 & 0 & 0 & 0 & 0 \\
0 & -e_{02} & 0 & 0 & 0 & 0 & 0 & 0 & a_{01} & 0 & 0 & 0 & 0 & 0 & 0 \\
-e_{10} & 0 & -e_{00} & 0 & 0 & 0 & a_{10} & 0 & 0 & a_{00} & 0 & 0 & 0 & 0 & 0 \\
-e_{11} & -e_{10} & -e_{01} & -e_{00} & 0 & 0 & a_{11} & a_{10} & 0 & a_{01} & a_{00} & 0 & 0 & 0 & 0 \\
-e_{12} & -e_{11} & -e_{02} & -e_{01} & 0 & 0 & 0 & a_{11} & a_{10} & 0 & a_{01} & a_{00} & 0 & 0 & 0 \\
0 & -e_{12} & 0 & -e_{02} & 0 & 0 & 0 & 0 & a_{11} & 0 & 0 & a_{01} & 0 & 0 & 0 \\
0 & 0 & -e_{10} & 0 & -e_{00} & 0 & 0 & 0 & 0 & a_{10} & 0 & 0 & a_{00} & 0 & 0 \\
0 & 0 & -e_{11} & -e_{10} & -e_{01} & -e_{00} & 0 & 0 & 0 & a_{11} & a_{10} & 0 & a_{01} & a_{00} & 0 \\
0 & 0 & -e_{12} & -e_{11} & -e_{02} & -e_{01} & 0 & 0 & 0 & 0 & a_{11} & a_{10} & 0 & a_{01} & a_{00} \\
0 & 0 & 0 & -e_{12} & 0 & -e_{02} & 0 & 0 & 0 & 0 & 0 & a_{11} & 0 & 0 & a_{01} \\
0 & 0 & 0 & 0 & -e_{10} & 0 & 0 & 0 & 0 & 0 & 0 & 0 & a_{10} & 0 & 0 \\
0 & 0 & 0 & 0 & -e_{11} & -e_{10} & 0 & 0 & 0 & 0 & 0 & 0 & a_{11} & a_{10} & 0 \\
0 & 0 & 0 & 0 & -e_{12} & -e_{11} & 0 & 0 & 0 & 0 & 0 & 0 & 0 & a_{11} & a_{10} \\
0 & 0 & 0 & 0 & 0 & -e_{12} & 0 & 0 & 0 & 0 & 0 & 0 & 0 & 0 & a_{11}
\end{bmatrix}
,(3.71b)
$$

the column c is the same as in (3.70) and the column b is given by

$$
b^T = \begin{bmatrix} f_{00} & f_{01} & f_{10} & f_{11} & f_{20} & f_{21} & b_{00} & b_{01} & b_{02} & b_{10} & b_{11} & b_{12} & b_{02} & b_{21} & b_{22} \end{bmatrix}.(3.78)
$$

The condition (3.72) can again be examined as in Method 2.

Example 3.13 Consider the polynomial of (3.66) characterized by the following coefficients, see (3.70),

$$
c^T = \begin{bmatrix} 1 & 0 & -2 & 3 & 2 & -1 & 2 & -3 & 2 & -3 & 0 & 2 & 1 & 2 & 0 & 3 \end{bmatrix}. \qquad (3.79)
$$

Now, we consider applying both the representations of (3.17) and (3.18) along with performing both the methods proposed previously.

Representation of (3.17) – the case of (3.67)

Method 1: The equations (3.68a) are solvable and have an infinite number of solutions. Two of these are (by way of example):

[a0 a00 a10 a20 a1 a21 a11 a2 a3 a01 b11 b2 b00 b10 b01 b3 b02

b12 b0 b1]

= [1.72188391511863 -1.59603722545499 -1.35693930465743

　　0.138472354144739 -0.474106100925967 0.137458385553639

　　-0.404110613855529 -1.65291820197556 -1.72976599182218

　　-0.517195094376272 0.985959596281266 -1.30305310778157

　　0.329963081606925 0.213271885946348 -0.0252611521736193

　　1.75942885213191 0.919447188907104 -0.315733504048298

　　0.274912014458419 -0.457713592341831]

and

= [-0.874718879964021 1.2934981928141 1.29867767234594

　　-0.5549561900063 0.0514658996055423 1.22273800495206

　　0.145480435800547 1.17741019265428 1.03265739744379

　　-0.605774112547794 -0.762951989883842 1.13411193646847

　　-1.02959588513112 0.336180609014633 -0.467091144489652

　　-3.21638705699298 -0.646588179967475 -0.262873894665643

　　0.379299477440622 1.03149046915933].

Both these solutions are not exact but approximate up to a prescribed level of accuracy. The values of d_i, i=1,2,...,16, when substituting these solutions, are

[-.631e-7, .712e-7, .53e-7, -.20e-7, -.31e-7, -.108e-6, .77e-7, -.27e-7, -.34e-7,

.24e-7, .8459e-7, -.69e-7, -.767e-7, -.102e-6, .1222e-6, -.454e-7]

and

[.3e-8, -.111e-7, .17e-7, -.6e-8, 0, .99e-8, -.20e-7, -.7e-8, .30e-7, .16e-7, .67e-8,

.13e-7, -.102e-7, -.9e-8, .301e-7, -.73e-8]

respectively.

Method 2: The number of equations to be solved has been reduced from 16 (Method 1) to 6, see (3.71)-(3.75). However, their form becomes very complicated and tedious.

The Maple V operations to derive the final equation set are

```
> A:=matrix(16,10, [a0,0,0,0,-a00,0,0,0,0,0,0,a0,0,0,0,
    -a10,-a00,0,0,0,0,0,0,a0,0,-a20,-a10,0,0,0,0,
    0,0,0,a0,0,-a20,0,0,0,0,0,0,a1,0,-a21,-a11,-a20,-
    a10,0,0,0,0,0,a1,0,-a21,0,-a20,0,0,0,0,a2,0,0,0,
    -a21,-a11,-a20,-a10,0,0,0,a2,0,0,0,-a21,0,-a20,
    0,0,a3,0,0,0,0,0,-a21,-a11,0,0,0,a3,0,0,0,0,0,
    -a21,a1,0,0,0,-a01,0,-a00,0,0,0,0,a1,0,0,-a11,-
    a01,-a10,-a00,0,0,a2,0,0,0,0,0,-a01,0,-a00,0,
    0,a2,0,0,0,0,-a11,-a01,-a10,-a00,a3,0,0,0,0,0,
    0,0,-a01,0,0,a3,0,0,0,0,0,0,-a11,-a01]):
> C:=matrix(16,1, [1,0,-2,3,2,-1,2,-3,2,-3,0,
    2,1,2,0,3]):
> A1:=LUdecomp(A,P='P1',L='L1'):
> evalm(L1):
> evalm(P1):
> L2:=inverse(L1):
> d:=map(simplify,evalm(L2&*C)):
> x1:=eval(d[11,1]):
> x2:=eval(d[12,1]):
> x3:=eval(d[13,1]):
> x4:=eval(d[14,1]):
> x5:=eval(d[15,1]):
> x6:=eval(d[16,1]):
> e1:=x1=0:
> e2:=x2=0:
> e3:=x3=0:
```

```
>  e4:=x4=0:
```

```
>  e5:=x5=0:
```

```
>  e6:=x6=0:
```

Application of Method 1 to the equation set e_i, i=1,2,...,6 gives an infinite number of solutions, such as (by way of example):

[a0 a00 a10 a20 a1 a21 a11 a2 a3 a01]

= [-1.9699393445076 -0.122269451483345 1.59832387018482

   -0.767155769864393 -2.71198032840423 -4.78679939200157

   2.75983798896605 -1.68387219100515 -2.56118086046535

   1.88904823881863]

or

= [-6.26361414328848 -1.1492786395832 0.493061348406336

   14.4418181023541 1.03062024582688 -15.3036923038912

   5.30718767340711 -8.54854710799884 7.40983050092329

   5.69798032990515].

Now, we can solve the equation set (3.75), where the solution is: for the first case

b^T = [-.4663658660 -.6113377381 2.214429910 -1.974194566 .6648229005

   -1.158882700 -.07272890932 -1.051465495 .6323011177

   -1.683017958],

and for the second

= [-.1640752994 -.7360091788e-1 .4072748458 .4697562132

   -.2410587360e-1 -.4114697761 .2762126673e-1 -.3332590500

   -.2133686127 -.4234804117].

For the accuracy of the solutions, we revisit the column c resulting from (3.71) when substituting on the left-hand side the above solution b for the concrete matrix X calculated for the given set of [a0 a00 a10 a20 a1 a21 a11 a2 a3 a01]. Hence, we have

[1.000000000 -.1e-9 -2.000000000 3.000000000 1.999999999

-.9999999983 2.000000001 -3.000000000 2.000000002 -3.000000002

.1149e-8 1.999999989 1.000000042 2.000000013 .17e-7 2.999999979],

and

[1.000000000 0 -1.999999999 2.999999999 2.000000001 -1.000000000

2.000000000 -3.000000000 2.000000001 -2.999999999 .6 10e-10

2.000000003 1.000000003 2.000000003 0 3.000000001],

respectively, which is a good approximation to (3.79).

Representation of (3.18) - the case of (3.76)

Both of the methods discussed previously can again be used. The first of these is used in much the same way as in the first case and hence is not discussed further here. Hence, consider the use of the second in the new framework. Then, applying the procedure (3.73)-(3.75) can again yield an infinite number of solutions. Here, two of these are presented:

a00=6.01215973064139, a01=-2.24340282898426, a10=-3.31799947064125,

a11=-0.990456748239858, e00=-1.880676899024, e01=0.470429128882692,

e02=-6.47279875545536, e10=-0.755033511506955,

e11=-1.31436710914912, e12=3.17682573861945

and

a00=-0.968609815995782, a01=0.283485043503517,

a10=-0.160768938424782, a11=0.332344091176045,

e00=0.0798989326973676, e01=-0.416966808694566,

e02=1.20126306648296, e10=0.621012028323284,

e11=-0.109752082484649, e12=0.7075375640419,

respectively. Deriving the rest of polynomial coefficients, *i.e.* solving (3.71) when the components are defined by (3.71b), and (3.78), is however somewhat more complicated than in the previous case. This is because the 16×15 matrix X is now not full-column rank and that its rank does not exceed 13. It is easy to see that the submatrix XX

    XX:=submatrix(X,1..13,[1,2,3,4,5,6,7,8,9,10,11,13,15])

is of a maximal rank. Thus, assuming zero values for b_{11}, b_{02}, and b_{22} the following operations:

```
>XX1:=subs(a00= , a01= , a10= , a11= , e00= , e01= ,

e02= , e10= , e11= , e12= , eval(XX));

>C1:=submatrix(C,1..13,[1]);

>x:=linsolve(AA1,C1);
```

where the values of elements calculated from the previous stage are substituted,
yield the required solution, *i.e.*

$$b^T = [-.8745164493 \quad .6778415471 \quad .1117794153 \quad -.03615493934$$

$$1.811518053 \quad -.3592171498 \quad .4398889920 \quad -.1163227147 \quad .6184943268$$

$$.6502854014 \quad .2107074839 \quad .1108369185 \quad -1.876739355 \,]$$

and

$$= [-15.43311868 \quad 2.270468488 \quad 6.171246978 \quad 6.749661305 \quad .2671258649$$

$$-7.054333461 \quad .2406435565 \quad -6.760500768 \quad 20.20364061$$

$$7.280938272 \quad 3.263441392 \quad -7.251950450 \quad -5.991396593],$$

respectively. Clearly, one can also assume arbitrary values for b_{11}, b_{02}, and b_{22} and
then update the right-hand side of (3.71), which leads to equivalent solutions.

It is also possible to use other initial representations than (3.17) and (3.18).
Hence, consider the existence of the representation

$$a(s_1,s_2) = \sum_{i=0}^{t_1}\sum_{j=0}^{t_2} a_{ij} s_1^{t_1-i} s_2^{t_2-j} = \det \begin{bmatrix} s_1 + \alpha & \sum\limits_{i=0}^{t_1}\sum\limits_{j=0}^{t_2-1} b_{ij} s_1^{t_1-i} s_2^{t_2-1-j} \\ s_2 + \beta & \sum\limits_{i=0}^{t_1-1}\sum\limits_{j=0}^{t_2} c_{ij} s_1^{t_1-1-i} s_2^{t_2-j} \end{bmatrix} \tag{3.80}$$

for a principal polynomial. Also, note that this representation is the subject of
Example 3.8.

Assuming the variables α and β to be parameters, the polynomial equation
(3.80) can be treated as a linear vector equation with notation

$$a_k = \begin{bmatrix} a_{k0} \\ a_{k1} \\ \cdots \\ a_{kt_1} \end{bmatrix}, c_k = \begin{bmatrix} c_{k0} \\ c_{k1} \\ \cdots \\ c_{kt_1-1} \end{bmatrix}, k = 0,1,...,t_2; b_k = \begin{bmatrix} b_{k0} \\ b_{k1} \\ \cdots \\ b_{kt_1} \end{bmatrix}, k = 0,1,...,t_2 - 1 \tag{3.81}$$

and

$$\Gamma_\alpha = \begin{bmatrix} 1 & 0 & & & 0 \\ \alpha & 1 & & & \\ 0 & \alpha & \ddots & & \\ & & \ddots & 1 & \\ 0 & & & & \alpha \end{bmatrix} \in \mathrm{Mat}_{t_1+1,t_1}(R);$$

$$\Gamma_\beta = \begin{bmatrix} 1 & 0 & & & 0 \\ \beta & 1 & & & \\ 0 & \beta & \ddots & & \\ & & \ddots & 1 & \\ 0 & & & & \beta \end{bmatrix} \in \mathrm{Mat}_{t_2+1,t_2}(R). \tag{3.82}$$

Now, (3.80) can be written as

$$\Theta X = \overline{A}, \tag{3.83}$$

where $\Theta = \begin{bmatrix} I_{t_2+1} \otimes \Gamma_\alpha & -\Gamma_\beta \otimes I_{t_1+1} \end{bmatrix}$, $\overline{A} = \begin{bmatrix} a_0^T & a_1^T & \cdots & a_{t_2}^T \end{bmatrix}^T$,

$$X = \begin{bmatrix} c_0^T & c_1^T & \cdots & c_{t_2}^T & b_0^T & b_1^T & \cdots & b_{t_2-1}^T \end{bmatrix}^T. \tag{3.84}$$

It is obvious that (3.83) has solution if and only if

$$\mathrm{rank}\,\Theta = \mathrm{rank}\begin{bmatrix} \Theta & \overline{A} \end{bmatrix}. \tag{3.85}$$

Applying appropriate elementary operations on the matrices Θ and $\begin{bmatrix} \Theta & \overline{A} \end{bmatrix}$ yields the equivalent condition

$$\det\begin{bmatrix} -\Gamma_\beta & \vdots & \begin{matrix} \gamma_0 \\ \gamma_1 \\ \cdots \\ \gamma_{t_2} \end{matrix} \end{bmatrix} = 0, \tag{3.86}$$

where

$$\gamma_k = \sum_{l=0}^{t_1}(-1)^l c_{lk}\alpha^{t_1-l}, k = 0,1,...,t_2. \tag{3.87}$$

Calculating the determinant (3.86) yields the polynomial equation

$$\sum_{k=0}^{t_2}(-1)^k \gamma_k \beta^{t_2-k} = 0. \tag{3.88}$$

Hence, we have to choose α and β in such a way that (3.88) holds. Then, (3.83) is solvable and its solution yields an appropriate representation (3.80). It is clear that such a representation always exists for a given monic bivariate polynomial, when the complex solution is admissible. However the EOA performed directly to the representation of (3.80) can yield only singular solutions, *cf.* Example 3.8. In Chapter 6 we describe the possibility of avoiding unnecessary singularity by using variable transforms such as inversion and bilinear transform.

3.5 Relationships between the Roesser and the Fornasini-Marchesini Models

Consider now a particular case of initial representations for the EOA, which are of the dimension greater than 2 and can provide a general result (without any rank limitation) in a very easy way. This is a square matrix, whose entries are first order in each variable, and:

all diagonal polynomials are principal,

all non-diagonal polynomials are not principal.

This ensures that the resulting characteristic matrix is nonsingular (of course, for a principal polynomial). For a monic polynomial, *i.e.* when the leading coefficient is unity, such a representation can be written as

$$\Psi(s_1, s_2) = Is_1 s_2 + A_1 s_1 + A_2 s_2 + A_3 . \tag{3.89}$$

Example 3.14 Given the polynomial

$$a(s_1, s_2) = s_1^4 s_2^4 + 6s_1^4 s_2^3 + 15s_1^4 s_2^2 + 27s_1^4 s_2 + 9s_1^4 - 3s_1^3 s_2^4 - 18s_1^3 s_2^3 - 39s_1^3 s_2^2$$

$$- 80s_1^3 s_2 + 57s_1^3 - s_1^2 s_2^4 - 32s_1^2 s_2^3 + 79s_1^2 s_2 - 127s_1^2 - 5s_1 s_2^4 + 41s_1 s_2^3$$

$$+ s_1 s_2^2 + 347s_1 s_2 - 115s_1 - 6s_2^4 + 23s_2^3 + 270s_2^2 - 457s_2 + 192$$

with the following initial representation

$$\Psi(s_1, s_2)$$

$$= \begin{bmatrix} s_1 s_2 + s_1 - s_2 + 1 & -s_1 - s_2 + 4 & 2s_1 + s_2 - 1 & -2s_1 + s_2 + 1 \\ s_1 - 2s_2 + 3 & s_1 s_2 + 2s_1 - 2s_2 + 1 & 3s_1 - s_2 + 2 & s_1 + s_2 - 2 \\ -s_1 + 2s_2 + 2 & s_1 + 3s_2 - 1 & s_1 s_2 + 2s_1 - s_2 + 1 & -s_1 - s_2 + 3 \\ s_1 + s_2 - 2 & -s_1 - 2s_2 + 2 & -s_1 - 3s_2 + 1 & s_1 s_2 + s_1 + s_2 + 2 \end{bmatrix}$$

apply the EOA procedure:

Augment($B_1(s_1,s_2) = \Psi(s_1,s_2)$)), $L(2+1\times(s_2+1))$, $L(3+1)$, $L(4+1\times(-1))$, $L(5+1)$ $R(2+1\times(-s_1+1)) \rightarrow B_2(s_1,s_2)$,

Augment($B_2(s_1,s_2)$)), $\quad$ $L(3+1\times(-1))$, $\quad$ $L(4+1\times(s_2+2))$, $\quad$ $L(5+1)$, $L(6+1\times(-1))$, $R(4+1\times(-s_1+2)) \rightarrow B_3(s_1,s_2)$,

Augment($B_3(s_1,s_2)$)), $\quad$ $L(4+1\times 2)$, $\quad$ $L(5+1\times 3)$, $\quad$ $L(6+1\times(s_2+2))$, $L(7+1\times(-1))$, $R(6+1\times(-s_1+1)) \rightarrow B_4(s_1,s_2)$,

Augment($B_4(s_1,s_2)$)), $\quad$ $L(5+1\times(-2))$, $\quad$ $L(6+1)$, $\quad$ $L(7+1\times(-1))$, $L(8+1\times(s_2+1))$ $R(8+1\times(-s_1-1)) \rightarrow B_5(s_1,s_2)$.

In the following, apply the subsequent non-augmenting operations:

$R(5+1\times(1))$, $R(6+1\times 2)$, $R(7+1\times 3)$, $R(5+2\times(-2))$, $R(6+2\times(-3))$, $R(8+2)$, $R(5+3)$, $R(7+3)$, $R(8+3\times(-1))$, $R(6+4)$, $R(7+4\times(-1))$, $R(8+4\times(-1))$.

Final multiplication of the first four rows by '-1' and performing suitable row and column permutations yield the characteristic matrix of the form

$$
\begin{bmatrix}
s_1+1 & -3 & -2 & 1 & 0 & 0 & 0 & -1 \\
-1 & s_1-1 & 3 & 2 & 0 & 0 & -1 & 0 \\
1 & -1 & s_1-2 & -1 & 0 & -1 & 0 & 0 \\
1 & 1 & -1 & s_1-1 & -1 & 0 & 0 & 0 \\
5 & -7 & -7 & -1 & s_2+1 & -1 & 2 & -2 \\
-3 & 9 & -1 & -1 & 1 & s_2+2 & 3 & 1 \\
6 & 2 & -8 & -1 & -1 & 1 & s_2+2 & -1 \\
0 & 1 & 6 & -1 & 1 & -1 & -1 & s_2+1
\end{bmatrix}.
$$

It is easy to see that the submatrix H_{12} is always of the full rank since it is the unit matrix with the sign '-1' and the rank of H_{21} can be of the full rank equal here 4, which depends on the polynomial coefficients. It is also easy to prepare an automated procedure, which produces the required result. Also, note that the representation (3.89) may be referred to the Fornasini-Marchesini state-space model, see (2.30). Note, however, that the initial representation of such a form refers only to the polynomials with equal degrees in both the variables. In the following, we generalize this result.

First, note that it is possible to use block elementary operations[*] defined similarly as for the scalar case:

L(i×c)   - multiplication of the ith block row by a real matrix c,

R(i×c)   - multiplication of the ith block column by c,

L(i+j×b(s_1,s_2))   - addition to the ith block row of the jth block row

   premultiplied by a polynomial matrix $b(s_1,s_2)$ of the respective size,

R(i+j×b(s_1,s_2))   - addition to the ith block column of the jth block column
   right multiplied by a polynomial matrix $b(s_1,s_2)$ of the respective size,

L(i,j) - interchange of the ith and jth block rows,

R(i,j) - interchange of the ith and jth block columns,

and the block augmenting operator

$$\textbf{BlockAugment}_\alpha : \text{Mat}_{n,n}(R[s_1,s_2]) \rightarrow \text{Mat}_{n+\alpha,n+\alpha}(R[s_1,s_2]), \tag{3.90}$$

given as

$$\textbf{BlockAugment}_\alpha (\Omega(s_1,s_2)) = \begin{bmatrix} I_\alpha & 0 \\ 0 & \Omega(s_1,\ s_2) \end{bmatrix}. \tag{3.91}$$

The initial representations related to the Fornasini-Marchesini form lead to the interesting conclusions on the equivalence of the both aforementioned models.

Theorem 3.5 *The characteristic matrix* $(s_1 s_2 I_n - A_1 s_1 - A_2 s_2 - A_3)$, *cf. (3.89), of the standard, i.e.* $E' = I_n$, *Fornasini-Marchesini model form, can be transformed by using the elementary operation approach to the Roesser model type*

$$\begin{bmatrix} I_n s_1 - A_2 & A_3 + A_2 A_1 \\ I_n & I_n s_2 - A_1 \end{bmatrix} \tag{3.92}$$

or

$$\begin{bmatrix} I_n s_1 - A_2 & I_n \\ A_3 + A_1 A_2 & I_n s_2 - A_1 \end{bmatrix}. \tag{3.93}$$

Proof Perform the operations

---

[*]The idea of the block notation of the elementary operations arose due to Kaczorek comments.

$$\text{BlockAugment}_n\left(s_1 s_2 I_n - A_1 s_1 - A_2 s_2 - A_3\right)$$

$$= \begin{bmatrix} I_n & 0 \\ 0 & s_1 s_2 I_n - A_1 s_1 - A_2 s_2 - A_3 \end{bmatrix}, \tag{3.94}$$

and next

$$\mathbf{L}\left(2 + 1 \times \left(I_n s_1 - A_2\right)\right), \mathbf{R}\left(2 + 1 \times \left(-I_n s_2 + A_1\right)\right)$$

or

$$\mathbf{L}\left(2 + 1 \times \left(I_n s_2 - A_1\right)\right), \mathbf{R}\left(2 + 1 \times \left(-I_n s_1 + A_2\right)\right). \tag{3.95}$$

(Note that each elementary operation is applied to the matrix resulting from the application of the previous one). Finally, applying appropriate block row and column permutation yields the required form of (3.92). ∎

This result is the same as before, where the elementary operations were applied in a non-block manner. Also, it is seen that elementary operations confirm, from the algebraic standpoint, links between the Roesser and Fornasini-Marchesini models. Conversely, from the EOA point of view, the rank condition (3.61) has been weakened since one of the non-diagonal blocks has the rank equal to n and the second depends on the initial model (3.59).

The previous result can also be generalized to the singular case; *i.e.* a characteristic matrix of the form $\left(s_1 s_2 E' - A_1 s_1 - A_2 s_2 - A_3\right)$ is to be considered.

Theorem 3.6 *The characteristic matrix* $\left(s_1 s_2 E' - A_1 s_1 - A_2 s_2 - A_3\right)$ *of the Fornasini-Marchesini form, can be transformed by using the elementary operation approach to the Roesser-type form*

$$\begin{bmatrix} E' s_1 - A_2 & -A_1 s_1 - A_3 \\ I_n & I_n s_2 \end{bmatrix} \tag{3.96}$$

or

$$\begin{bmatrix} I_n s_1 & -A_2 s_2 - A_3 \\ I_n & -E' s_2 - A_1 \end{bmatrix}. \tag{3.97}$$

Proof Perform the operations

$$\text{BlockAugment}_n\left(s_1 s_2 E' - A_1 s_1 - A_2 s_2 - A_3\right)$$

$$= \begin{bmatrix} I_n & 0 \\ 0 & s_1 s_2 E' - A_1 s_1 - A_2 s_2 - A_3 \end{bmatrix}, \tag{3.98}$$

and next

$$\mathbf{L}\left(2+1\times\left(E'_{s_1}-A_2\right)\right), \mathbf{R}\left(2+1\times\left(-I_n s_2\right)\right)$$

or

$$\mathbf{L}\left(2+1\times I_n s_1\right), \mathbf{R}\left(2+1\times\left(E'_{s_2}-A_1\right)\right) \tag{3.99}$$

and finally appropriate block and row permutations.　■

The first result (3.96) represents the following singular Roesser model

$$\begin{bmatrix} E' & -A_1 \\ 0 & -I_n \end{bmatrix}\begin{bmatrix} \tilde{\xi}(i+1,j) \\ x(i,j+1) \end{bmatrix} = \begin{bmatrix} A_2 & A_3 \\ -I_n & 0 \end{bmatrix}\begin{bmatrix} \tilde{\xi}(i,j) \\ x(i,j) \end{bmatrix} + \begin{bmatrix} B \\ 0 \end{bmatrix} u(i,j) \tag{3.100}$$

where

$$\tilde{\xi}(i,j) = x(i,j+1). \tag{3.101}$$

when (3.97) yields the model shown in (2.10).

In turn, this result is examined in the case in which, however, non-block approach is applied.

Example 3.15 Consider the non-principal polynomial

$$a(s_1,s_2) = s_1^3 s_2^2 + s_1^2 s_2^3 - 2s_1^3 s_2 - 2s_1 s_2^3 - s_2^3 + 6s_1^2 s_2^2 + 2s_1^2 s_2 - 11 s_1 s_2^2$$

$$+ 4s_1^2 - 8s_2^2 - 6s_1 s_2,$$

which can be represented as the determinant of the following polynomial matrix:

$$\Psi(s_1,s_2) = \begin{bmatrix} s_1 s_2 + 2s_1 + 1 & s_1 + 1 & s_1 + s_2 - 1 \\ s_2 + 1 & s_1 s_2 - s_1 + s_2 - 1 & s_1 - 2s_2 + 1 \\ 2s_1 - s_2 + 3 & s_1 - 2s_2 - 1 & s_1 + s_2 + 1 \end{bmatrix}.$$

Now, apply the EOA procedure:

Augment($B_1(s_1,s_2) = \Psi(s_1,s_2)$),　L(2+1×$s_1$),　L(3+1),　L(4+1×(−1)),
R(2+1×(−s_2−2)),　R(3+1×(−1)),　L(4+1×(−1))

and the following non-augmenting operation

$$R(4+1\times(-1)) \rightarrow B_2(s_1,s_2),$$

Augment($B_2(s_1,s_2)$),　L(6+1×s_1)),　R(4+1×(−2)),　R(5+1×(−1)),
R(6+1×(−1)) → $B_3(s_1,s_2)$,

Augment($B_3(s_1,s_2)$),　L(5+1×(−1)),　L(6+1×2),　L(7+1×(−1)),
R(7+1×s_2) → $B_4(s_1,s_2)$.

Finally, multiplication of the third and fourth columns by '-1' and applying suitable row and column permutations yield the characteristic matrix of the form

$$
\left[
\begin{array}{ccc|ccc}
0 & 1 & 0 & 0 & 2 & -1 & -1 \\
\hline
-1 & s_1 & -2 & -1 & -5 & -2 & -2 \\
2 & 0 & s_1+1 & 1 & 1 & 1 & -1 \\
-1 & 0 & 0 & s_1 & -1 & 1 & -1 \\
\hline
0 & 0 & 0 & 1 & s_2+2 & -1 & -1 \\
0 & 0 & -1 & 0 & 0 & s_2-1 & 1 \\
1 & 0 & 0 & 0 & 0 & 0 & s_2
\end{array}
\right].
$$

It is seen that this method gives the result of the form where the matrix $E\Lambda$ satisfies

$$
E\Lambda = \begin{bmatrix} 0 & 0 & 0 \\ 0 & s_1 I_{\tau_1} & 0 \\ 0 & 0 & s_2 I_{\tau_2} \end{bmatrix}, \tag{3.102}
$$

which confirms the results of Section 3.3.

The results of Theorems 3.5 and 3.6 are well-known, see e.g. Kaczorek (1998), Kung (1985). These also generalize and lead to very interesting results reported e.g. in Gałkowski (1996b,c). Introduce the following polynomial matrix:

$$
\Phi(s_1,s_2) = \begin{bmatrix} I_n s_1 s_2 - A_{11}^1 s_1 - A_{11}^2 s_2 - A_{11}^3 & -A_{12}^3 & -A_{13}^3 \\ -A_{21}^3 & I_{k_1} s_1 - A_{22}^3 & -A_{23}^3 \\ -A_{31}^3 & -A_{32}^3 & I_{k_2} s_2 - A_{33}^3 \end{bmatrix}, \tag{3.103}
$$

which is clearly related to the singular general or Fornasini-Marchesini model, where

$$
E = \begin{bmatrix} I_n & 0 \\ 0 & 0 \end{bmatrix}, \quad A_1 = \begin{bmatrix} A_{11}^1 & 0 & 0 \\ 0 & -I_{k_1} & 0 \\ 0 & 0 & 0 \end{bmatrix}, \quad A_2 = \begin{bmatrix} A_{11}^2 & 0 & 0 \\ 0 & 0 & 0 \\ 0 & 0 & -I_{k_2} \end{bmatrix}. \tag{3.104}
$$

and A_3 is an arbitrary matrix of the respective dimensions.

Theorem 3.7 *The matrix* $\Phi(s_1,s_2)$ *of (3.103) can be transformed by using the elementary operation approach to the Roesser type form*

$$
\Lambda - H = \begin{bmatrix} I_n s_1 - A_{11}^2 & -A_{12}^3 & A_{11}^3 + A_{11}^2 A_{11}^1 & -A_{13}^3 \\ 0 & I_{k_1} s_1 - A_{22}^3 & A_{21}^3 & -A_{23}^3 \\ I_n & 0 & I_n s_2 - A_{11}^1 & 0 \\ 0 & -A_{32}^3 & A_{31}^3 & I_{k_2} s_2 - A_{33}^3 \end{bmatrix}, \tag{3.105}
$$

where

$$t_1 = n + k_1, t_2 = n + k_2.$$ (3.106)

Proof Perform the operations

$$\mathbf{BlockAugment}_n(\Phi(s_1, s_2)) = \begin{bmatrix} I_n & 0 \\ 0 & \Phi(s_1, s_2) \end{bmatrix},$$

$$\mathbf{L}\left(2 + 1 \times \left(I_n s_1 - A_{11}^2\right)\right) \mathbf{R}\left(2 + 1 \times \left(-I_n s_2 + A_{11}^1\right)\right)$$ (3.107)

Finally, applying the appropriate block row and column permutation yields the required form of (3.105). ∎

It is clear that the rank of H_{21} can vary from n to $\min(t_1, t_2)$ and the rank of H_{12} is arbitrary or vice versa. The matrix (3.105) can be considered as the most general result since we have obtained an arbitrary 2D companion matrix H without any rank limitations, and what is more in a comparatively easy manner. Moreover, this result shows explicitly that the singular Fornasini-Marchesini model can be related to the standard Roesser one.

Similar results can also be obtained for other initial representations of the type considered previously, but this leads to the singular case.

Theorem 3.8 *The characteristic matrix of the form*

$$\Phi(s_1, s_2) = \begin{bmatrix} I_n s_1 s_2 - A_{11}^1 s_1 - A_{11}^2 s_2 - A_{11}^3 & -A_{12}^3 & -A_{13}^3 \\ A_{21}^2 s_2 - A_{21}^3 & I_{k_1} s_1 - A_{22}^3 & -A_{23}^3 \\ A_{31}^1 s_1 - A_{31}^3 & -A_{32}^3 & I_{k_2} s_2 - A_{33}^3 \end{bmatrix}$$ (3.108)

can be transformed by using the elementary operation approach to the markedly singular characteristic matrix of the Roesser type

$$E'\Lambda' - H'$$

$$= \begin{bmatrix} 0 & 0 & 0 & I_n & A_{31}^1 & 0 \\ 0 & I_n s_1 - A_{11}^2 & -A_{12}^3 & 0 & A_{11}^3 + A_{11}^2 A_{11}^1 & -A_{13}^3 \\ 0 & A_{21}^2 & I_{k_1} s_1 - A_{22}^3 & 0 & A_{21}^3 - A_{21}^2 A_1 & -A_{23}^3 \\ I_n & 0 & 0 & I_n s_1 & 0 & 0 \\ 0 & I_n & 0 & 0 & I_n s_2 - A_{11}^1 & 0 \\ -I_n & 0 & -A_{32}^3 & 0 & A_{31}^3 & I_{k_2} s_2 - A_{33}^3 \end{bmatrix},$$ (3.109)

where

$$E' = \begin{bmatrix} 0_{n,n} & 0 \\ 0 & I_{3n+k_1+k_2} \end{bmatrix}, \Lambda' = s_1 I_{3n+k_1} \oplus s_2 I_{n+k_2}.$$ (3.110)

Proof Perform the operations:

BlockAugment$_n$($\Phi(s_1,s_2)$),

$$L\left(2+1\times\left(I_n s_1 - A_{11}^2\right)\right), L\left(3+1\times A_{21}^2\right), R\left(2+1\times\left(-I_n s_2 + A_{11}^1\right)\right) \Rightarrow \Phi_1(s_1,s_2),$$

BlockAugment$_n$ $\left(\Phi_1(s_1,s_2)\right), L\left(5+1\times I_n s_1\right), R\left(3+1\times\left(-A_{31}^1\right)\right) \Rightarrow \Phi_2(s_1,s_2),$

BlockAugment$_n$ $\left(\Phi_2(s_1,s_2)\right), L\left(6+1\times\left(-I_n\right)\right), R\left(2+1\times I_n s_1\right) \Rightarrow \Phi_3(s_1,s_2).$

Finally, applying appropriate column and row permutations and multiplying some of them by "-1", yields the required matrix of (3.109). ∎

A similar result can be obtained e.g. for the matrix

$$\Phi(s_1,s_2) = \begin{bmatrix} I_n s_1 s_2 - A_{11}^1 s_1 - A_{11}^2 s_2 - A_{11}^3 & A_{12}^2 s_2 - A_{12}^3 & -A_{13}^3 \\ -A_{21}^3 & I_{k_1} s_1 - A_{22}^3 & -A_{23}^3 \\ A_{31}^1 s_1 - A_{31}^3 & -A_{32}^3 & I_{k_2} s_2 - A_{33}^3 \end{bmatrix}. \quad (3.111)$$

4. 2D State-space Realizations and the Elementary Operation Algorithm – Single Input - Single Output (SISO) Case

Previously, it has been shown how to use the EOA procedure to obtain a characteristic matrix for a given bivariate polynomial in the standard or singular form. Three cases have been considered. The first of these virtually uses the given polynomial as the starting point. In the second, the 2×2 matrix of (3.17), or (3.18) is used with further modifications e.g. (3.62). In the third method, the starting point is a system matrix of the Fornasini-Marchesini type in the form of (3.89).

Here, we generalise the method to deal with bivariate rational functions, which clearly represent the transfer functions for linear 2D SISO systems. The crucial point here is again finding a characteristic matrix $\Lambda-H$ or $E\Lambda-H$ for a given bivariate polynomial, which then provides the state transition matrix A in the Roesser model (2.1) or (2.5), respectively. This is, however, only a part of the solution, since we have not considered the transfer function numerator and therefore we do not obtain the rest of the model matrices. To remove this difficulty, we can convert a given bivariate transfer function into the equivalent trivariate polynomial form, cf. (2.45)-(2.46). As shown in Chapter 2, the characteristic matrix $\Lambda-H$ (or $E\Lambda-H$) for such a trivariate polynomial provides a full state-space realization for a given bivariate transfer function. Hence, we can apply the EOA to the trivariate polynomial representation of a 2D transfer function, which solves the problem of finding its state-space realization.

In this chapter, we revisit the previously developed procedures and apply them to trivariate polynomials/polynomial matrices to develop a new effective methodology for the construction of state–space realizations for 2D transfer functions. This, in turn, leads to using the EOA to provide an easy method of linking the Fornasini-Marchesini and Roesser models and hence to some generalizations of existing results.

4.1 Basics

The basics of the method are first detailed for a standard 1D SISO system. Hence, noting (2.41)-(2.42), start with the polynomial

$$A_f(s, z) = \left[a_f(s, z) \right] = \left[za(s) - b(s) \right] \tag{4.1}$$

treated as a square 1×1 polynomial matrix and perform the EOA procedure. First, augment the matrix size by using the operator Augment, *i.e.*

$$\text{Augment}\big(A_f(s,z)\big) = \begin{bmatrix} 1 & 0 \\ 0 & za(s) - b(s) \end{bmatrix}. \tag{4.2}$$

Next, successively apply the following chain of elementary operations:

$$L(2+1\times z), R(2+1\times(-a(s))). \tag{4.3}$$

The final result is

$$A_{f_1}(s,z) = \begin{bmatrix} 1 & -a(s) \\ \hline z & -b(s) \end{bmatrix}. \tag{4.4}$$

At this stage the polynomials $a(s)$ and $b(s)$ can be written as

$$a(s) = (s+\alpha_1)a_1(s) + x_1, \quad b(s) = (s+\alpha_1)b_1(s) + y_1, \tag{4.5}$$

where x_1, y_1 and $\alpha_1 \in R$. Also, the term $(s+\alpha_1)$ can obviously be chosen in such a way that $a_1(s)$ and $b_1(s)$ are polynomials with the degree one less than $a(s)$ and $b(s)$, respectively. This step of the algorithm ends by applying the operations

$$\text{Augment}\big(A_{f_1}(s,z)\big), \ L(2+1\times a_1(s)), \ L(3+1\times b_1(s)), \ R(3+1\times(s+\alpha_1)), \tag{4.6}$$

which yields

$$A_{f_2} = \begin{bmatrix} 1 & 0 & s+\alpha_1 \\ a_1(s) & 1 & -x_1 \\ b_1(s) & z & -y_1 \end{bmatrix}. \tag{4.7}$$

Now, write the polynomials $a_1(s)$ and $b_1(s)$ in the same form as $a(s)$ and $b(s)$, *i.e.*

$$a_1(s) = (s+\alpha_2)a_2(s) + x_2, \quad b_1(s) = (s+\alpha_2)b_2(s) + y_2. \tag{4.8}$$

where x_2, y_2 and $\alpha_2 \in R$. Then, as before, the term $(s+\alpha_2)$ can be chosen such that $a_2(s)$ and $b_2(s)$ are polynomials with the degree one less than $a_1(s)$ and $b_1(s)$ respectively. Now, apply the operations

$$\text{Augment}\big(A_{f_2}(s,z)\big), \ L(3+1\times(-a_2(s))), \ L(4+1\times(-b_2(s))),$$
$$R(4+1\times(s+\alpha_2)), \tag{4.9}$$

which yields

$$A_{f_3}(s,z) = \begin{bmatrix} 1 & s+\alpha_2 & 0 & 0 \\ 0 & 1 & 0 & s+\alpha_1 \\ -a_2(s) & x_2 & 1 & -x_1 \\ -b_2(s) & y_2 & z & -y_1 \end{bmatrix}.$$ (4.10)

It is obvious that applying this procedure a suitable number of times (equal to the maximum degree of the polynomials $a(s)$ and $b(s)$) yields

$$A_{f_t}(s,z) = \begin{bmatrix} 1 & s+\alpha_{t-1} & \cdots & 0 & 0 & 0 \\ 0 & 1 & \ddots & 0 & 0 & 0 \\ \vdots & \vdots & \ddots & \vdots & \vdots & \vdots \\ 0 & 0 & \cdots & s+\alpha_2 & 0 & 0 \\ 0 & 0 & \cdots & 1 & 0 & s+\alpha_1 \\ \varepsilon_1 s+\phi_1 & (+/-)x_{t-1} & \cdots & x_2 & 1 & -x_1 \\ \varepsilon_2 s+\phi_2 & (+/-)y_{t-1} & \cdots & y_2 & z & -y_1 \end{bmatrix}.$$ (4.11)

Also, if the degree of the transfer function denominator is greater than that of the numerator, then

$$\varepsilon_2 = 0.$$ (4.12)

If, however, the degrees are equal then

$$\varepsilon_2 \neq 0$$ (4.13)

and the following additional operation is needed:

$$L\left((t+1) + t\times\left(-\frac{\varepsilon_2}{\varepsilon_1}\right)\right).$$ (4.14)

Also, note that the resulting matrix has the following property: Each column and each row have only one monomial element of the type $\varepsilon s+\phi$ and the remaining elements are real numbers. Hence, there exist row and column permutations, and row and column multiplications by real numbers such that the system matrix

$$A_f(s,z) = \begin{bmatrix} sI_t - A & -B \\ -C & z-D \end{bmatrix}$$ (4.15)

is obtained. Here, it is easy to see how the method leads to an efficient implementation algorithm for producing the required result from the initial representation. Note also that it is clearly possible to develop other EOA operation sequences that produce essentially different results.

Consider again 2D transfer functions and associate the following 1×1 polynomial matrix with such a transfer function

$$A_f(s_1,s_2,z) = [a_f(s_1,s_2,z)] = [za(s_1,s_2) - b(s_1,s_2)]$$ (4.16)

and use virtually the same approach as shown before for the 1D case. In particular, first augment the matrix size by using the operator Augment, *i.e.*

$$\text{Augment}\left(A_f(s_1,s_2,z)\right)=\begin{bmatrix} 1 & 0 \\ 0 & za(s_1,s_2)-b(s_1,s_2) \end{bmatrix}. \tag{4.17}$$

Next, successively apply the following chain of elementary operations:

$$L(2+1\times z), R(2+1\times(-a(s_1,s_2,))). \tag{4.18}$$

The result of this stage is

$$A_{f_1}(s,z)=\begin{bmatrix} 1 & \vdots & -a(s_1,s_2) \\ \cdots & \cdots & \cdots \\ z & \vdots & -b(s_1,s_2) \end{bmatrix}. \tag{4.19}$$

Now, apply the EOA:

a. to augment step by step the polynomial matrix and

b. to lower the degrees of polynomial entries of the matrix, and to partition their variables by using appropriate elementary polynomial operations.

Hence, the resulting final matrix has the following property: Each column and each row have only one monomial element of the type $\varepsilon s_i+\phi, i=1,2$ and the remaining elements are real numbers. Finally, there exist row and column permutations, and row and column multiplications by real numbers, such that the system matrix related to the Roesser form

$$\mathbf{A}_f(s_1,s_2,z)=\begin{bmatrix} s_1 I_{t_1}\oplus s_2 I_{t_2}-A & -B \\ -C & z-D \end{bmatrix} \tag{4.20}$$

is obtained. It is not very instructive to present the general steps of the algorithm for bivariate transfer functions and instead we demonstrate the method by means of appropriate examples in the Maple environment for scientific computations. These are given next together with the Maple commands and some auxiliary procedures.

Example 4.1 Consider the bivariate transfer function

$$f(s_1,s_2)=\frac{2s_1^2 s_2-3s_1 s_2+s_2-1}{s_1^2 s_2+2s_1 s_2-s_2+1}.$$

The associated polynomial is

$$a_f(s_1,s_2,s_3)=s_3\left(s_1^2 s_2+2s_1 s_2-s_2+1\right)-\left(2s_1^2 s_2-3s_1 s_2+s_2-1\right).$$

Apply now the chain of elementary operations and size augmentations shown below:

```
> B1:=Augment(a_f);
> B1:=addcol(B1,1,2,-z);
> B1:=map(simplify,addrow(B1,1,2,s^2*p+2*s*p-p+1));
```

```
> B2:=Augment(B1);

> B2:=addrow(B2,1,3,p);

> B2:=map(simplify,addcol(B2,1,2,-s^2.2*s+1));

> B2:=map(simplify,addcol(B2,1,3,2*s^2.3*s+1));

> B2:=addcol(B2,2,3,2);

> B3:=Augment(B2);

> B3:=addrow(B3,1,2,s);

> B3:=map(simplify,addcol(B3,1,3,s+2));

> B3:=addcol(B3,1,4,7);

> B3:=mulrow(B3,3,-1);
```

where, for the notational simplicity the variables have been renamed as

$$s_1 \Rightarrow s, s_2 \Rightarrow p, s_3 \Rightarrow z \tag{4.21}$$

and apply the appropriate row and column permutations to obtain the following system matrix of the Roesser type:

$$\Lambda - H = \begin{bmatrix} s_1 + 2 & 1 & 0 & 7 \\ 1 & s_1 & 1 & 3 \\ 1 & 0 & s_2 & 3 \\ -1 & 0 & 0 & s_3 - 2 \end{bmatrix}.$$

This matrix obviously represents the Roesser type state-space realization

$$A = \begin{bmatrix} -1 & -1 & 0 \\ -1 & 0 & -1 \\ -1 & 0 & 0 \end{bmatrix}, \ b = \begin{bmatrix} -5 \\ -3 \\ -3 \end{bmatrix}, \ c = \begin{bmatrix} 1 & 0 & 0 \end{bmatrix}, \ d = 2$$

where $t_1 = 2, t_2 = 1$. In this example two special procedures have been used, *i.e.* for matrix augmentation:

```
> Augment := proc(MAT::matrix)
>        local M, w, k;
>        w := linalg[rowdim](MAT);
>        k := linalg[coldim](MAT);
>        M := matrix(w + 1, k + 1, 0);
>        M[1,1] := 1;
>        linalg[copyinto](MAT,M,2,2);
>        RETURN(evalm(M));

> end:
```

and for the row and columns permutations (here, the procedure parameters (r, c, A) have been given according to the problem solved)

```
> with(group):
> with(linalg):
> r:=4:                    #row number of matrices A and AA
> c:=4:                    #column number of matrices A and AA
> A:=array(1..r,1..c):     #original matrix
> AA:=array(1..r,1..c):    #permuted matrix
> A:=B3:
> pr:=[1,2,4,3]:           #permutation for rows
> pc:=[3,1,2,4]:  #permutation for columns
> permat:=proc(r, c, A, AA, pr, pc)
>     local i, k;
>       for k from 1 to c do
>          for i from 1 to r do
>             AA[i,k]:=A[pr[i],pc[k]]:
>          od:
>       od:
>     print(AA):
> end:
```

Note that applying the above procedure to an arbitrary transfer function very often yields a singular final realisation irregardless of whether the associated polynomial a_f is principal or non-principal. The following example demonstrates this point.

Example 4.2 Consider the bivariate transfer function

$$f(s_1,s_2) = \frac{s_1 + s_2 - 1}{s_1^2 s_2 - 2s_1 s_2 - s_2 + 1}.$$

The associated polynomial is

$$a_f(s_1,s_2,s_3) = s_3(s_1^2 s_2 - 2s_1 s_2 - s_2 + 1) - (s_1 + s_2 - 1).$$

Apply now the chain of elementary operations and size augmentations shown below. Here, again notational simplicity, the variables have been renamed as in (4.21).

```
> B1:=Augment(a_f);
> B1:=addcol(B1,1,2,-z);
```

```
> B1:=map(simplify,addrow(B1,1,2,s^2*p-2*s*p+3*s*p-
p+1));
> B2:=Augment(B1);
> B2:=addrow(B2,1,3,p);
> B2:=map(simplify,addcol(B2,1,2,-s^2.s+1));
> B2:=map(simplify,addcol(B2,1,3,1));
> B3:=Augment(B2);
> B3:=addrow(B3,1,4,s);
> B3:=addcol(B3,1,4,1);
> B3:=mulrow(B3,3,-1);
> B4:=Augment(B3);
> B4:=addrow(B4,1,3,s+1);
> B4:=map(simplify,addcol(B4,1,4,s));
> B5:=Augment(B4);
> B5:=addcol(B5,1,3,s);
> B5:=addrow(B5,1,6,-1);
> permat(r,c,A,AA,pr,pc);
```

In this case, the operation "permat" implements the following row and column permutations:

```
> pr:=[3,1,2,4,6,5]:      #permutation for rows
> pc:=[1,3,5,2,4,6]: #permutation for columns
```

which results in the system matrix

$$
E\Lambda - H = \begin{bmatrix}
0 & 1 & 0 & 0 & 0 & 1 \\
1 & s_1 & 0 & 0 & 0 & 0 \\
0 & 0 & s_1 & 1 & 0 & 0 \\
0 & 0 & 1 & s_1+1 & 1 & 1 \\
-1 & 0 & 1 & 0 & s_2 & 1 \\
0 & 0 & -1 & 0 & 0 & s_3
\end{bmatrix},
$$

clearly singular even though the transfer function is principal. It is also easy to see that the singularity property of the final result depends on the mutual co-existence of the transfer function numerator and denominator. This disadvantage can be prevented to some extent by using different initial representations or, if this is impossible, by using variable transformations, where this last approach is treated further in a subsequent chapter here.

Now, the matrix representations of the polynomial $a_f(s_1,s_2,s_3)$, similar to those developed for bivariate polynomials, are assumed to be the starting point for applying the EOA. Start, therefore, with a 3D polynomial matrix $\Xi(s_1,s_2,s_3)$ of the form

$$\Xi(s_1,s_2,s_3) = \left[\begin{array}{c|c} \Psi(s_1,s_2) & \begin{matrix} * \left(s_1^k\right) \\ * \left(s_2^l\right) \end{matrix} \\ \hline * \left(s_1^{t_1-k}\right) \ * \left(s_2^{t_2-l}\right) & s_3 - \alpha \end{array} \right] \tag{4.22}$$

where the 2×2 block submatrix $\Psi(s_1,s_2)$ represents the transfer function denominator and hence can be defined as in (3.17) or (3.18) or as discussed in Section 3.4. Moreover, it is straightforward that the numerator can be written as

$$a(s_1,s_2) = -\chi(s_1,s_2) + \alpha b(s_1,s_2), \tag{4.23}$$

where

$$\chi(s_1,s_2) = \det \Xi_0(s_1,s_2) = \det \left[\begin{array}{c|c} \Psi(s_1,s_2) & \begin{matrix} * \left(s_1^k\right) \\ * \left(s_2^l\right) \end{matrix} \\ \hline * \left(s_1^{t_1-k}\right) \ * \left(s_2^{t_2-l}\right) & 0 \end{array} \right]. \tag{4.24}$$

Now, the following essential theorem can be established.

Theorem 4.1 *Applying the EOA to the polynomial matrix $\Xi(s_1,s_2,s_3)$ of (4.22) for a given transfer function, i.e. such that*

$$a_f(s_1,s_2,s_3) \triangleq \det \Xi(s_1,s_2,s_3), \tag{4.25}$$

yields the Roesser type standard system matrix $\lambda - H$ where

$$\lambda = s_1 I_{t_1} \oplus s_2 I_{t_2} \oplus s_3. \tag{4.26}$$

Proof The proof is immediate on application of the elementary operations of the form given in Section 3.1. ∎

Since performing the EOA in this case is based on virtually the same procedures as for bivariate polynomials, which has been discussed previously, there is no need to present the general procedure again. However, some examples are given next to illustrate the application of the method.

Example 4.3 Consider the transfer function

$$f(s_1,s_2) = \frac{s_2^2 \left(4s_1^2 + 47s_1 + 24\right) - s_2 \left(5s_1^2 - 34s_1 - 1\right) - 2s_1^2 + 16s_1 - 14}{s_2^2 \left(s_1^2 + 9s_1\right) - s_2 \left(3s_1^2 - 5s_1 + 3\right) - s_1^2 + 6s_1 + 3}.$$

Then the trivariate polynomial $a_f(s_1, s_2, s_3)$ is given as

$$a_f(s_1, s_2, s_3) = s_3\left(s_2^2\left(s_1^2 + 9s_1\right) - s_2\left(3s_1^2 - 5s_1 + 3\right) - s_1^2 + 6s_1 + 3\right)$$

$$-\left(s_2^2\left(4s_1^2 + 47s_1 + 24\right) - s_2\left(5s_1^2 - 34s_1 - 1\right) - 2s_1^2 + 16s_1 - 14\right)$$

It is easy to see that this polynomial can be written as

$$a_f(s_1, s_2, s_3) = \det\begin{bmatrix} s_1^2 + 3s_1 + 1 & s_1 s_2 + s_1 - s_2 - 2 & s_1 + 4 \\ s_1 s_2 + 2s_1 - 2s_2 + 1 & 2s_2^2 + 1 & 4s_2 + 3 \\ 2s_1 - 1 & 3s_2 + 1 & s_3 + 1 \end{bmatrix},$$

which is consistent with (4.22) and (3.17). Now, apply the following sequence of operations:

1. Augment$(\Xi(s_1, s_2, s_3))$,

2. $L(2 + 1 \times (s_1 - 1))$, $L(3 + 1 \times (s_2 + 2))$, $L(4 + 1 \times 2)$, $R(2 + 1 \times (-s_1 + 2))$, $R(3 + 1 \times (-s_2 - 1))$, $R(4 + 1 \times (-1)) \Rightarrow \Xi_1(s_1, s_2, s_3)$,

3. Augment$(\Xi_1(s_1, s_2, s_3))$,

4. $L(2 + 1 \times 1)$, $L(4 + 1 \times (-s_2 + 3))$, $L(5 + 1 \times (-1))$, $R(4 + 1 \times s_2)$, $R(5 + 1 \times 3)$, $L(3 + 2 \times 6)$, $R(2 + 1 \times 1) \Rightarrow \Xi_2(s_1, s_2, s_3)$.

Finally, applying appropriate row and column permutations to $\Xi_2(s_1, s_2, s_3)$ and multiplications by "-1" yields

$$\lambda - H = \begin{bmatrix} s_1 - 2 & 2 & -1 & -1 & 2 \\ -11 & s_1 + 11 & -6 & -7 & 17 \\ -5 & 5 & s_2 - 3 & -1 & 10 \\ 0 & 1 & -1 & s_2 & 3 \\ -3 & 1 & 1 & -1 & s_3 - 4 \end{bmatrix}.$$

Hence, the nonsingular Roesser model (2.1), (2.2) where

$$A = \begin{bmatrix} 2 & -2 & 1 & 1 \\ 11 & -11 & 6 & 7 \\ 5 & -5 & 3 & 1 \\ 0 & -1 & 1 & 0 \end{bmatrix}, B = \begin{bmatrix} -2 \\ -17 \\ -10 \\ -3 \end{bmatrix}, C = \begin{bmatrix} 3 & -1 & -1 & 1 \end{bmatrix}, D = 4$$

has been obtained.

To discuss the EOA is able to produce various, equivalent state-space realizations for a system described by its transfer function, return to Example 4.1 and consider the same transfer function, but now represented in the form of (4.22).

Example 4.4 It is easy to see that the associated polynomial a_f from Example 4.1 can be represented as

$$a_f(s_1,s_2,s_3) = \det \begin{bmatrix} s_1{}^2 + 2s_1 - 1 & 1 & 7 \\ -1 & s_2 & 0 \\ -s_1 + \dfrac{3}{7} & -\dfrac{3}{7} & s_3 - 2 \end{bmatrix}.$$

Now, apply matrix the following sequence of operations:

```
> B1:=Augment( A_f(s_1,s_2,s_3));
> B1:=addcol(B1,1,2,-s);
> B1:=map(simplify,addrow(B1,1,2,s+2));
> B1:=addrow(B1,1,4,-1);
> B1:=mulrow(B1,1,-1);
> permat(r,c,A,AA,pr,pc);
```

which yields the following 3D system matrix of the Roesser type

$$\Lambda - H' = \begin{bmatrix} s_1 & -1 & 0 & 0 \\ -1 & s_1+2 & 1 & 7 \\ -1 & 0 & s_2 & 0 \\ \dfrac{3}{7} & -1 & -\dfrac{3}{7} & s_3-2 \end{bmatrix},$$

which obviously represents the Roesser type state-space realization

$$A' = \begin{bmatrix} 0 & 1 & 0 \\ 1 & -2 & -1 \\ 1 & 0 & 0 \end{bmatrix}, \quad b' = \begin{bmatrix} 0 \\ -7 \\ 0 \end{bmatrix}, \quad c' = \begin{bmatrix} -\dfrac{3}{7} & 1 & \dfrac{3}{7} \end{bmatrix}, \quad d' = 2$$

where $t_1 = 2, t_2 = 1$. Here, again for the notational simplicity the variables have been renamed as in (4.21). Moreover, it is easy to see that both the matrices H, from Example 4.1, and H' are similar with the similarity matrix T given (as one possibility) by

$$T = \begin{bmatrix} 0 & 0 & 0 & 1 \\ 1 & 0 & 0 & 2 \\ 0 & 0 & -1 & -1 \\ -\dfrac{2}{7} & \dfrac{1}{7} & \dfrac{1}{7} & \dfrac{1}{7} \end{bmatrix},$$

where $H = T^{-1}HT$. Of particular note here is that, this matrix is again not block diagonal with respect to the matrix $\Lambda = s_1 I_2 \oplus s_2 \oplus s_3$ (compare to the analysis for the bivariate polynomials in Section 3.2). Also, the state transition matrices A

and A', which are clearly sub-matrices of H and H', are also similar, *i.e.*
$A' = R^{-1}AR$, where (as one possibility)

$$R = \begin{bmatrix} 1 & 0 & 0 \\ -2 & -1 & 0 \\ 0 & 0 & 1 \end{bmatrix}.$$

Hence, the matrix R is not a submatrix of T and in this particular case R is even block diagonal with respect to $\Lambda' = s_1 I_2 \oplus s_2$.

Theorem 4.1 provides, in fact, the sufficient condition for the existence of a minimal state-space realization of the Roesser form for a 2D system described by a rational bivariate transfer function when it is co-prime. Hence, the problem of how to derive the representation (4.22) is extremely important. First, the analysis of (4.23)-(4.24), establishes that the matrix Ψ has to be the companion matrix for the denominator polynomial. However it does not always provide the possibility of partitioning the problem in the sense that, given a transfer function f, we produce an arbitrary companion matrix for its denominator polynomial and next using (4.23)-(4.24) we derive the rest of polynomials in (4.22) to achieve the prescribed numerator polynomial. Unfortunately, choosing a particular initial representation Ψ for a transfer function denominator may influence the solvability of the resulting equation set. Hence, for some initial representations Ψ the solution may exist but for others it may not.

The problem of constructing the representation Ψ for a given polynomial a_f is very similar to that presented for a bivariate polynomial in Section 3.4 but now we have to solve the trilinear equations set. Hence, we can proceed in a similar manner to before but only using the first of the two methods detailed in Section 3.4. Note here that the second method from Section 3.4 does not easily generalize to this case due to the fact that the equation set is trilinear. One possible option here would be to make use of the Grobner basis based techniques but this is still very much an open research question.

In what follows, the direct method for calculation of the representation $\Xi(s_1, s_2, s_3)$ of (4.22) for a given transfer function is discussed. Consider for simplicity the particular polynomial

$$a_f(s_1, s_2, s_3) = \sum_{i_1=0}^{2} \sum_{i_2=0}^{2} \sum_{i_3=0}^{1} a_{i_1 i_2 i_3} s_1^{2-i_1} s_2^{2-i_2} s_3^{1-i_3}, \tag{4.27}$$

which clearly represents the transfer function

$$f(s_1, s_2) = -\frac{\displaystyle\sum_{i_1=0}^{2} \sum_{i_2=0}^{2} a_{i_1 i_2 1} s_1^{2-i_1} s_2^{2-i_2}}{\displaystyle\sum_{i_1=0}^{2} \sum_{i_2=0}^{2} a_{i_1 i_2 0} s_1^{2-i_1} s_2^{2-i_2}}. \tag{4.28}$$

Note that the polynomial $\Xi(s_1, s_2)$ of (4.22), when the initial representation $\Psi(s_1, s_2)$ has the form of (3.17), can be rewritten as

$$\Xi(s_1, s_2, s_3)$$

$$= \left[\begin{array}{cc|c} a_0 s_1^2 + a_1 s_1 + a_2 & b_{00} s_1 s_2 + b_{01} s_1 + b_{10} s_2 + b_{11} & c_0 s_1 + c_1 \\ a_{00} s_1 s_2 + a_{01} s_1 + a_{10} s_2 + a_{11} & b_0 s_2^2 + b_1 s_2 + b_2 & d_0 s_2 + d_1 \\ \hline e_0 s_1 + e_1 & f_0 s_2 + f_1 & s_3 - g \end{array} \right]. \quad (4.29)$$

Now, the requirement to make $\det \Xi$ equal to the prescribed polynomial (4.27) can be rewritten in the form of the following trilinear equations:

$$y_i = 0, \quad x_i = 0, \quad i = 1, 2, \ldots, 9, \quad (4.30)$$

where the variables $y_i(x_i)$ relate to a transfer function denominator (numerator), and

```
> y1:=a0*b0-a00*b00-a000(1):

> y2:=a1*b0-a01*b00-a00*b01-a010(0):

> y3:=a0*b1-a10*b00-a00*b10-a100(5):

> y4:=a1*b1-a11*b00-a01*b10-a10*b01-a00*b11-a110(2):

> y5:=a0*b2.a01*b01-a020(-2):

> y6:=a2*b0-a10*b10-a200(-2):

> y7:=a1*b2.a11*b01-a01*b11-a120(2):

> y8:=a2*b1-a11*b10-a10*b11-a210(-1):

> y9:=a2*b2.a11*b11-a220(2):

> x1:=-g*a0*b0+g*b00*a00-a0*d0*f0+b00*d0*e0-c0*b0*e0
+c0*a00*f0-a001(3):

> x2:=g*b00*a01+c0*a01*f0+c0*a00*f1-c0*e0*b1-a0*d0*f1
-g*a0*b1+b01*d0*e0-a0*d1*f0+b00*e0*d1+g*b01*a00
-a011   (-1):

> x3:=-a1*d0*f0+g*b10*a00+g*b00*a10+b00*d0*e1-g*a1*b0
-c1*b0*e0-c0*b0*e1+c0*a10*f0+c1*a00*f0+b10*d0*e0
-a101 (-2):

> x4:=-g*a0*b2+g*b01*a01+b01*e0*d1-c0*e0*b2+c0*a01*f1
-a0*f1*d1-a021(-1):

> x5:=b01*d0*e1+g*b10*a01*g*a1*b1+g*b00*a11+b11*d0*e0
+c0*a10*f1+c0*a11*f0+g*b11*a00+c1*a00*f1-c1*e0*b1
```

```
+c1*a01*f0-c0*b1*e1-a1*d1*f0-a1*d0*f1+b10*e0*d1
+b00*d1*e1+g*b01*a10-a111(-3):
> x6:=-g*a2*b0+g*b10*a10+b10*d0*e1+c1*a10*f0-a2*d0*f0
-c1*b0*e1-a201(0):
> x7:=-g*a1*b2+g*b01*a11+b11*e0*d1-a1*f1*d1+c0*a11*f1
-c0*b2*e1-c1*e0*b2+g*b11*a01+c1*a01*f1+b01*d1*e1
-a121(-4):
> x8:=g*b11*a10-g*a2*b1-a2*d0*f1+b11*d0*e1-a2*d1*f0
+c1*a10*f1+b10*d1*e1-c1*b1*e1+g*b10*a11+c1*a11*f0
-a211(1):
> x9:=g*b11*a11-a2*f1*d1-g*a2*b2+b11*d1*e1+c1*a11*f1-
c1*b2*e1-a221(-1):                                         (4.31)
```

The equation set (4.30)-(4.31) can be solved numerically using a wide variety of methods, see e.g. Ortega, Rheinboldt (2000).

Example 4.5 Consider the equation set (4.31) for a particular choice of the coefficients a_{ijk} ($i, j = 0,1,2; k = 0,1$). These are given in the right hand side brackets of the equations of (4.31). In fact, infinitely many solutions exist and two of these now follow. Firstly

$a0 = .840476208523722$, $a1 = -.88782865697736$, $a2 = -.558075541351595$,

$b00 = -2.1835760006793$, $b01 = .04063715510777334$,

$b10 = 1.05587016292706$, $b11 = -.521371055035185$, $c0 = 2.01586585423615$,

$c1 = .630556426686209$, $a00 = .588170692553563$, $a01 = -.12659585703567$,

$a10 = 2.07296800342528$, $a11 = 1.28236033297417$, $b0 = -.338279012610385$,

$b1 = 1.30229613010748$, $b2 = -2.38572454163094$, $d0 = -1.65430751218737$,

$d1 = -1.74062738661931$, $e0 = .447742474428625$, $e1 = -1.31682251629134$,

$f0 = .0244443533993515$, $f1 = -.904999195270259$, $g = -1.01431956383545$

and secondly

$a0 = -1.77041798814131$, $a1 = .170140310295387$, $a2 = .751960212612178$,

$b00 = -1.18398182646295$, $b01 = .771817053761939$,

b10 = 1.07872617700355, b11 = 1.27236649707941, c0 = 1.30729716631417,

c1 = 1.06329684989158, a00 = -1.88605506615858,

a01 = -.967065106178902, a10 = .581059864747097,

a11 = -.655084251876504, b0 = -1.82615361743217,

b1 = -1.28641873036574, b2 = 1.55127024796442, d0 = .250961106733061,

d1 = -.378720931323856, e0 = .89810341096464, e1 = -1.04200814666877,

f0 = -1.25942977002589, f1 = .910724513661548, g = 1.42292933734098.

Substitution these solutions into the equation set gives in the first case

$[-.529 \ 10^{-7}(\approx 0), .54490 \ 10^{-6}, .211 \ 10^{-6}, -.127 \ 10^{-6}, -.223 \ 10^{-6}, .180 \ 10^{-6},$

$-.271 \ 10^{-6}, -.117 \ 10^{-6}, .75 \ 10^{-6}, -.153 \ 10^{-7}, -.275 \ 10^{-6}, -.20 \ 10^{-7}, .2538 \ 10^{-6},$

$.425 \ 10^{-6}, .1805 \ 10^{-6}, -.77 \ 10^{-7}, .206 \ 10^{-6}, -.15 \ 10^{-7}]$

and in the second case

$[.291 \ 10^{-6} (\approx 0), -.389 \ 10^{-6}, .151 \ 10^{-6}, .279 \ 10^{-6}, .590 \ 10^{-6}, .651 \ 10^{-6},$

$.41 \ 10^{-7}, -.276 \ 10^{-6}, .760 \ 10^{-6}, .9 \ 10^{-8}, .58 \ 10^{-6}, -.436 \ 10^{-6}, .759 \ 10^{-7},$

$.390 \ 10^{-6}, .117 \ 10^{-6}, -.229 \ 10^{-6}, .16 \ 10^{-7}, .826 \ 10^{-6}],$

which confirms that the solution is proper.

Also, this proves that the zero coprime transfer function

$$f(s_1, s_2) = \frac{-3s_1^2 s_2^2 + s_1^2 s_2 + 2 s_1 s_2^2 + s_1^2 + 3 s_1 s_2 + 4 s_1 - s_2 + 1}{s_1^2 s_2^2 + 5 s_1 s_2^2 + 2 s_1 s_2 - 2 s_1^2 - 2 s_2^2 + 2 s_1 - s_2 + 2}$$

possesses a variety of state-space realizations of the dimension '4', which are, in fact, absolutely minimal, *i.e.* the state-space dimension is equal to the global denominator degree. It is related to the existence of initial representations of (4.22), (3.17), which enables the production of a corresponding EOA based algorithm.

Example 4.6 Consider the transfer function

$$f(s_1, s_2) = \frac{s_1^2 s_2^2 - 5 s_1^2 s_2 - 8 s_1 s_2^2 + 30 s_1^2 + 33 s_1 s_2 + 3 s_2^2 - 55 s_1 + 2 s_2 - 25}{s_1^2 s_2^2 - 2 s_1^2 s_2 - 3 s_1 s_2 - 5 s_1^2 + s_2^2 + 5 s_1 - 3 s_2},$$

which is also zero coprime. The numerical algorithm used in the previous example for solving the equation set of the form (4.30)-(4.31) fails now. This does not establish that a minimal realization does not exist. In particular, there is still the possibility that alternative algorithm could find a solution. Note also that early attempts at Grobner bases based solutions here were not very successful because the resulting Maple based computations could not be effectively run on the available computing machines. This general approach does, however, have much potential, which is an obvious area for further research.

The major problem to be removed for effective application of the EOA to the SISO case is how to examine the existence of particular polynomial matrix representations for a transfer function such that application the EOA guarantees the existence of a realization and, ideally, that this realization is minimal. Even if the initial representation of the form of (4.22), (3.17) does not exist, the one determined by (4.22), (3.18) or some other may exist. Return, therefore, to Example 4.6, where the equation set of the type (4.30)-(4.31) for an initial representation

$$A_f\left(s_1, s_2, s_3\right)$$
$$= \begin{bmatrix} a_{00}s_1s_2 + a_{01}s_1 + a_{10}s_2 + a_{11} & b_{00}s_1s_2 + b_{01}s_1 + b_{10}s_2 + b_{11} & c_0s_1 + c_1 \\ c_{00}s_1s_2 + c_{01}s_1 + c_{10}s_2 + c_{11} & d_{00}s_1s_2 + d_{01}s_1 + d_{10}s_2 + d_{11} & d_0s_1 + d_1 \\ \hline e_0s_2 + e_1 & f_0s_2 + f_1 & s_3 - g \end{bmatrix}, \quad (4.32)$$

is

```
-c00*b00 + a00*d00 - a000 (1) = 0,

-c00*b01 - c01*b00 + a01*d00 + a00*d01 - a010 (-2)

= 0,

-c10*b00 + a10*d00 - c00*b10 + a00*d10 -a100 (0) = 0,

-a00*d00*g - e0*c0*d00 + c00*c0*f0 + c00*b00*g

+ e0*b00*d0 - a00*d0*f0 - a001 (-1) = 0,

a01*d01 - c01*b01 - a020 (-5) = 0,

-c11*b00 - c10*b01 - c00*b11 - c01*b10 + a11*d00

+ a00*d11 + a01*d10 + a10*d01 - a110 (-3) = 0,

-c10*b10 + a10*d10 - a200 (1) = 0,

c00*c0*f1 + c00*b01*g - a00*d0*f1 - a01*d00*g

+ e1*b00*d0 - a00*d01*g + c01*c0*f0 - a01*d0*f0

+ c01*b00*g + e0*b01*d0 - e0*c0*d01 - e1*c0*d00

- a011 (5) = 0,

-e0*c1*d00 + c10*b00*g + e0*b00*d1 + c00*c1*f0

- a10*d00*g - a00*d1*f0 - a00*d10*g - a10*d0*f0

+ c10*c0*f0 + c00*b10*g + e0*b10*d0 - e0*c0*d10
```

$-$ a101 (8) = 0,

a01*d11 $-$ c11*b01 $-$ c01*b11 + a11*d01 $-$ a120 (5) = 0,

a11*d10 $-$ c10*b11 $-$ c11*b10 + a10*d11 $-$ a210 (-3) = 0,

$-$a01*d01*g + c01*b01*g $-$ e1*c0*d01 + e1*b01*d0

+ c01*c0*f1 $-$ a01*d0*f1 $-$a021 (-30) = 0,

$-$ a11*d00*g $-$ a11*d0*f0 $-$ e1*c0*d10 $-$ a01*d10*g

$-$ e1*c1*d00 $-$ a10*d0*f1 $-$ a01*d1*f0 $-$ a10*d01*g

$-$ a00*d1*f1 $-$ e0*c1*d01 $-$ a00*d11*g $-$ e0*c0*d11

+ c00*c1*f1 + c10*b01*g + e1*b10*d0 + e1*b00*d1

+ e0*b11*d0 + e0*b01*d1 + c01*c1*f0 + c01*b10*g

+ c10*c0*f1 + c11*b00*g + c00*b11*g + c11*c0*f0

$-$ a111 (-33) = 0,

c10*c1*f0 $-$ e0*c1*d10 $-$ a10*d10*g + c10*b10*g

$-$ a10*d1*f0 + e0*b10*d1 $-$ a201 (-3) = 0,

$-$c11*b11 + a11*d11 $-$ a220 (0) = 0,

$-$a01*d11*g + e1*b11*d0 + c01*c1*f1 + c11*c0*f1

+ c01*b11*g + e1*b01*d1 $-$ e1*c1*d01 $-$ a11*d0*f1

$-$ a01*d1*f1 $-$ e1*c0*d11 + c11*b01*g $-$ a11*d01*g

$-$ a121 (55) = 0,

$-$e1*c1*d10 + c11*c1*f0 + e1*b10*d1 $-$ a10*d1*f1

$-$ e0*c1*d11 $-$ a10*d11*g + e0*b11*d1 $-$ a11*d10*g

+ c10*c1*f1 $-$ a11*d1*f0 + c10*b11*g + c11*b10*g

$-$ a211 (-2) = 0,

c11*b11*g + e1*b11*d1 + c11*c1*f1 $-$ a11*d11*g

$-$ a11*d1*f1 $-$ e1*c1*d11$-$ a221 (25) = 0.   (4.33)

Example 4.7 First, note that the numbers in brackets denote the transfer function coefficients, which are clearly the same as in Example 4.6. Now, the solution procedure that previously has failed succeeds and we obtain, for example, the following solution

g = -2.93478634634214, a00 = .583558313601343,

a01 = .0732538661674488, a10 = .0275762177124039,

a11 = 1.19038558055939, b00 = -.422997729595313,

b01 = -1.2546241714269, b10 = -1.30078056905125,

b11 = 1.73203420022063, c0 = 1.17227169069679, c1 = -3.21921824764575,

c00 = .411228335757415, c01 = -3.88769045604096,

c10 = .801942235609646, c11 = .187748462994751,

d00 = 1.41553796227748, d01 = -1.67103679336306,

d10 = -1.56483895185186, d11 = .273177800618588,

d0 = 1.69660275378624, d1 = -1.99292149690205, e0 = 1.45407656108939,

e1 = -7.28719451908482, f0 = .941661633776842, f1 = 3.53772902764324

which is a good approximation to the exact solution since its substitution into the original equation set gives

-.24017 10^{-5}, -.1011 10^{-5}, -.20234 10^{-5}, .8217 10^{-6}, -.321 10^{-6}, -.679 10^{-6},

-.14616 10^{-5}, .341 10^{-6}, -.43 10^{-7}, -.185 10^{-6}, -.340 10^{-6}, .10 10^{-6}, .10 10^{-6},

-.618 10^{-6}, .1560 10^{-6}, .5 10^{-7}, -.77 10^{-7}, -.4 10^{-7}.

In the case, when the minimal realization does not exist or is difficult to construct, one can look for a priori non-minimal realizations. In the following example, we consider such a realization for the transfer function investigated in previous two examples.

Example 4.8 Multiply the numerator and the denominator of the transfer function of Example 6 by s_1 to obtain

$$f(s_1, s_2) = \frac{s_1(-3s_1^2 s_2^2 + s_1^2 s_2 + 2s_1 s_2^2 + s_1^2 + 3s_1 s_2 + 4s_1 - s_2 + 1)}{s_1(s_1^2 s_2^2 + 5s_1 s_2^2 + 2s_1 s_2 - 2s_1^2 - 2s_2^2 + 2s_1 - s_2 + 2)}.$$

The initial representation of the form (4.29) is now

$A_f(s_1, s_2, s_3)$

$$
= \begin{bmatrix}
a_0 s_1^3 + a_1 s_1^2 + a_2 s_1 + a_3 & \begin{array}{c} b_{00} s_1^2 s_2 + b_{01} s_1^2 + b_{10} s_1 s_2 + b_{11} s_1 \\ b_{10} s_1 s_2 + b_{11} s_1 + b_{20} s_2 + b_{21} \end{array} & c_0 s_1 + c_1 \\
\hline
a_{00} s_1 s_2 + a_{01} s_1 + a_{10} s_2 + a_{11} & b_0 s_2^2 + b_1 s_2 + b_2 & d_0 s_2^2 + d_1 s_2 + d_3 \\
e_0 s_1 + e_1 & f_0 s_2 + f_1 & s_3 - g
\end{bmatrix} . \quad (4.34)
$$

The equation set of the form (4.30)-(4.31) here becomes

-a00 b00 + a0 b0 - 1 = 0,

-a01 b00 + a0 b1 - a00 b01 + 2 = 0,

-a10 b00 - a00 b10 + a1 b0 = 0,

a00 c0 f0 - a0 b0 g + a00 b00 g - a0 d0 f0 + e0 b00 d0

- e0 c0 b0 + 1 = 0,

a0 b2 - a01 b01 + 5 = 0,

a1 b1 - a10 b01 - a01 b10 - a00 b11 - a11 b00 + 3 = 0,

a2 b0 - a10 b10 - a00 b20 - 1 = 0,

-a0 b1 g - a0 d1 f0 + a00 b01 g - a0 d0 f1 + a00 c0 f1

+ a01 c0 f0 + e0 b01 d0 - e0 c0 b1 + e0 b00 d1

+ a01 b00 g - 5 = 0,

a10 c0 f0 + e1 b00 d0 - e0 c1 b0 + a10 b00 g

- a1 d0 f0 - e1 c0 b0 + a00 b10 g + a00 c1 f0

- a1 b0 g + e0 b10 d0 - 8 = 0,

- a11 b01 + a1 b2 - a01 b11 - 5 = 0,

- a00 b21 + a2 b1 - a11 b10 - a10 b11 - a01 b20 + 3

= 0,

a3 b0 - a10 b20 = 0,

a01 b01 g + a01 c0 f1 - a0 d1 f1 + e0 b01 d1

- e0 c0 b2 - a0 b2 g + 30 = 0,

a11 b00 g + e1 b00 d1 - e0 c1 b1 + a11 c0 f0

+ e1 b01 d0 - a1 d1 f0 - a1 d0 f1 - e1 c0 b1 - a1 b1 g

+ a10 c0 f1 + a01 b10 g + a01 c1 f0 + a00 c1 f1

+ a10 b01 g + e0 b10 d1 + a00 b11 g + e0 b11 d0 + 33

$= 0,$

$-e0\ c2\ b0 - e1\ c1\ b0 + e1\ b10\ d0 + a10\ c1\ f0$

$+ a10\ b10\ g + e0\ b20\ d0 - a2\ d0\ f0 + a00\ b20\ g$

$+ a00\ c2\ f0 - a2\ b0\ g + 3 = 0,$

$a2\ b2 - a01\ b21 - a11\ b11, -a10\ b21 - a11\ b20 + a3\ b1$

$= 0,$

$a01\ b11\ g + a11\ c0\ f1 - e1\ c0\ b2 + e0\ b11\ d1$

$+ e1\ b01\ d1 + a11\ b01\ g - a1\ d1\ f1 + a01\ c1\ f1$

$- a1\ b2\ g - e0\ c1\ b2 - 55 = 0,$

$-a2\ d1\ f0 + e1\ b10\ d1 + a11\ b10\ g - a2\ d0\ f1$

$+ a00\ c2\ f1 - e1\ c1\ b1 - e0\ c2\ b1 + a11\ c1\ f0$

$+ e1\ b11\ d0 + a01\ b20\ g - a2\ b1\ g + e0\ b21\ d0$

$+ a01\ c2\ f0 + a00\ b21\ g + e0\ b20\ d1 + a10\ c1\ f1$

$+ a10\ b11\ g + 2 = 0,$

$-e1\ c2\ b0 - a3\ d0\ f0 + e1\ b20\ d0 + a10\ b20\ g - a3\ b0\ g$

$+ a10\ c2\ f0 = 0,$

$-a11\ b21 + a3\ b2, -a2\ d1\ f1 - e1\ c1\ b2 + a01\ b21\ g$

$+ e1\ b11\ d1 - a2\ b2\ g + e0\ b21\ d1 + a11\ c1\ f1$

$+ a01\ c2\ f1 - e0\ c2\ b2 + a11\ b11\ g - 25 = 0,$

$a10\ b21\ g - a3\ d0\ f1 + a11\ b20\ g + e1\ b21\ d0$

$+ a10\ c2\ f1 - a3\ d1\ f0 - a3\ b1\ g - e1\ c2\ b1$

$+ a11\ c2\ f0 + e1\ b20\ d1 = 0,$

$a11\ c2\ f1 + e1\ b21\ d1 - a3\ d1\ f1 - a3\ b2\ g - e1\ c2\ b2$

$+ a11\ b21\ g = 0.$

In this case the numerical algorithm employed succeeds and we obtain, for example, the following solution:

b0 = .592782585105715, b1 = -1.90853866442155, b2 = -1.38877088192159,

d0 = -1.52369040505989, d1 = 2.44810714995049, e0 = -1.6444863035178, e1 = 4.45533411669635, f0 = -.40202419119358, f1 = -.282740023953755,

g = -2.40569594056302, a0 = 1.74622058676852, a1 = -000247755291724074,

a2 = 1.6866617067388, a3 = -.816818910688199 10^{-7} = 0,

b00 = -.72269652621669, b01 = 3.51773091139414, b10 = .127199648991869,

b11 = 1.37235632394105, b20 = .641161901571473 10^{-8} = 0,

b21 = -.178551812813191 10^{-7} = 0, c0 = 2.01473628483008,

c1 = 1.3289374999419, c2 = .118016467317343 10^{-7} = 0,

a00 = -.0203968925064495, a01 = .731978015673361,

a10 = -.00142201911511832, a11 = -1.7068360133604.

Substituting these solutions to the underlying equations shows that the solution is very accurate since

-.85019 10^{-5}, -.2742 10^{-5}, -.20948433 10^{-5}, .1764 10^{-5}, -.1997 10^{-5}, -.881 10^{-6},

.4567 10^{-5}, .553 10^{-6}, .247 10^{-6}, -.781 10^{-6}, .1373 10^{-5}, -.4841048510 10^{-7},

-.23 10^{-6}, .51 10^{-6}, .317 10^{-6}, .543 10^{-6}, .1668112391 10^{-6}, -.16 10^{-6}, .69 10^{-7},

-.1411133378 10^{-6}, .8296156547 10^{-7}, -.76 10^{-6}, .6556927661 10^{-6},

.2701035303 10^{-7}.

4.2 Links Between the Roesser and Fornasini-Marchesini Models by the EOA

Start from the trivariate system matrix

$$\Phi = \begin{bmatrix} Es_1s_2 - A_1s_1 - A_2s_2 - A_3 & -B_1s_1 - B_2s_2 - B \\ -C & s_3 - D \end{bmatrix}, \qquad (4.35)$$

which is clearly associated to the Fornasini-Marchesini models.

Theorem 4.2 *Applying the EOA to the 3D standard system matrix of the Fornasini-Marchesini type (4.35), where $E = I_n$, yields the singular 3D system matrix of the Roesser form*

$$
\begin{bmatrix}
0 & 0 & 0 & I_n & \vdots & B_2 \\
-I_n & I_n s_1 - A_2 & A_3 + A_2 A_1 & 0 & \vdots & -B_3 - A_2 B_1 \\
0 & I_n & I_n s_2 - A_1 & 0 & \vdots & B_1 \\
I_n & 0 & 0 & I_n s_2 & \vdots & 0 \\
\hline
0 & 0 & C & 0 & \vdots & s_3 - D
\end{bmatrix} .
\tag{4.36}
$$

Proof Apply the following operations to the standard system matrix.

a. **BlockAugment$_n\Phi$,**

$$
\mathbf{L}\big(2+1\times(I_n s_1 - A_2)\big),\ \mathbf{R}\big(2+1\times(-I_n s_2 + A_1)\big),\ \mathbf{R}(2+1\times B_1) \to \Phi_1 ,
$$

b. **BlockAugment$_n\Phi_1$,** $\mathbf{L}\big(3+1\times(I_n s_2)\big),\ \mathbf{R}(4+1\times B_2) \to \Phi_2 ,$

c. **BlockAugment$_n\Phi_2$,** $\mathbf{L}\big(4+1\times(-I_n)\big),\ \mathbf{R}(2+1\times I_n s_2).$

After the last step, performing appropriate row and column permutations and multiplications by "-1" yields the required final form of (4.36). Here again bold letters denote that the underlying procedures/operations are applied to block matrices. ∎

Hence, the singular Roesser model (2.1), (2.2), where all matrices are replaced by their barred equivalents, has been derived

$$
\overline{E} = \begin{bmatrix} 0_{n,n} & 0 \\ 0 & I_{3n} \end{bmatrix},\
\overline{A} = \begin{bmatrix}
0 & 0 & 0 & -I_n \\
I_n & A_2 & -A_3 - A_2 A_1 & 0 \\
0 & -I_n & A_1 & 0 \\
-I_n & 0 & 0 & 0
\end{bmatrix},\
\overline{B} = \begin{bmatrix} -B_2 \\ B_3 + A_2 B_1 \\ -B_1 \\ 0 \end{bmatrix},
$$

$$
\tag{4.37}
$$

$$
\overline{C} = \begin{bmatrix} 0 & 0 & -C & 0 \end{bmatrix},\ \overline{D} = D.
$$

Note that singularity in the obtained Roesser model can be removed when using other initial representations.

Theorem 4.3 *The 3D polynomial matrix of the form*

$$
\Psi(s_1, s_2, s_3) = \begin{bmatrix}
I_n s_1 s_2 - A_{11}^1 s_1 - A_{11}^2 s_2 - A_{11}^3 & -A_{12}^1 s_1 - A_{12}^3 \\
-A_{21}^2 s_2 - A_{21}^3 & s_3 - A_{22}^3
\end{bmatrix},
\tag{4.38}
$$

related also to the Fornasini-Marchesini model, can be transformed to the standard Roesser form by using the EOA.

Proof Apply the following chain of operations:

• **BlockAugment$_n\big(\Psi(s_1, s_2, s_3)\big),$**

$$\mathbf{L}\left(2+1\times\left(I_n s_1 - A_{11}^2\right)\right), \mathbf{L}\left(2+1\times\left(-A_{21}^2\right)\right),$$

•

$$\mathbf{R}\left(2+1\times\left(-I_n s_2 + A_{11}^1\right)\right), \mathbf{R}\left(2+1\times\left(-A_{12}^1\right)\right)$$

Finally, performing the appropriate row and column permutations and multiplications by "-1" yields

$$[\lambda - H] = \begin{bmatrix} I_n s_1 - A_{11}^2 & A_{11}^3 + A_{11}^2 A_{11}^1 & -A_{12}^3 - A_{11}^2 A_{12}^1 \\ I_n & I_n s_2 - A_{11}^1 & A_{12}^1 \\ -A_{21}^2 & A_{21}^3 + A_{21}^2 A_{11}^1 & s_3 - A_{22}^3 - A_{21}^2 A_{12}^1 \end{bmatrix}. \tag{4.39}$$

Hence, the following full state-space realization of the Roesser type

$$A = \begin{bmatrix} A_2 & -A_{11}^3 - A_{11}^2 A_{11}^1 \\ -I_n & A_{11}^1 \end{bmatrix}, B = \begin{bmatrix} A_{12}^3 + A_{11}^2 A_{12}^1 \\ -A_{12}^1 \end{bmatrix},$$

$$\tag{4.40}$$

$$C = \begin{bmatrix} A_{21}^2 & -A_{21}^3 - A_{21}^2 A_{11}^1 \end{bmatrix}, D = A_{22}^3 + A_{21}^2 A_{12}^1.$$

is obtained.   ∎

It is clearly seen that (4.38) is related to some generalization of the second-order Fornasini-Marchesini model ($A_{21}^2 = 0$, $A_{12}^1 = 0$). Note, however, that the form of (4.38) constitutes an interesting case of singularity. A preliminary inspection suggests that it is the standard case but, in fact, immediately reveals the state equation is standard, *i.e.*

$$x(i+1, j+1)$$
$$= A_{11}^1 x(i+1, j) + A_{11}^2 x(i, j+1) + A_{11}^3 x(i, j) + A_{12}^3 u(i, j) + A_{11}^1 u(i+1, j) \tag{4.41}$$

but the output equation is in fact singular, since it can be written as

$$y(i, j) = A_{21}^3 x(i, j) + A_{21}^2 x(i, j+1) + A_{22}^3 u(i, j). \tag{4.42}$$

The dual form is also possible by shifting the input vector in (4.41) in 'j', and in (4.42) the state vector is shifted in 'i'. This case is associated with the following system matrix:

$$\Psi(s_1, s_2, s_3) = \begin{bmatrix} I_n s_1 s_2 - A_{11}^1 s_1 - A_{11}^2 s_2 - A_{11}^3 & -A_{12}^2 s_2 - A_{12}^3 \\ -A_{21}^1 s_1 - A_{21}^3 & s_3 - A_{22}^3 \end{bmatrix}. \tag{4.43}$$

Theorem 4.4 *The 3D polynomial matrix of the form (4.43) related to the Fornasini-Marchesini model can be transformed to the standard Roesser form by using the EOA.*

Proof Apply the following chain of operations:

- **BlockAugment**$_n$ $(\Psi(s_1, s_2, s_3))$,

$$\mathbf{R}\big(2+1\times\big(I_n s_1 - A_{11}^2\big)\big), \mathbf{L}\big(2+1\times\big(A_{21}^1\big)\big),$$

-

$$\mathbf{L}\big(2+1\times\big(-I_n s_2 + A_{11}^1\big)\big), \mathbf{R}\big(2+1\times\big(-A_{12}^2\big)\big).$$

Finally, performing appropriate row and column permutations and multiplications by "-1" yields

$$[\lambda - H] = \begin{bmatrix} I_n s_1 - A_{11}^2 & I_n & -A_{12}^2 \\ A_{11}^3 + A_{11}^1 A_{11}^2 & I_n s_2 - A_{11}^1 & A_{12}^3 + A_{11}^1 A_{12}^2 \\ A_{21}^3 + A_{21}^1 A_{11}^2 & A_{21}^1 & s_3 - A_{22}^3 - A_{21}^2 A_{12}^1 \end{bmatrix}. \tag{4.44}$$

∎

Hence, we have shown that some singularities in the Fornasini-Marchesini model are avoided by transforming that to the Roesser case.

Based on these last results, a new 2D linear systems state-space model is defined. In particular, return to the most general system matrix of the form of (3.103). Then this matrix can be embedded in the following trivariate system matrix

$$\Theta(s_1, s_2)$$

$$= \begin{bmatrix} I_n s_1 s_2 - A_{11}^1 s_1 - A_{11}^2 s_2 - A_{11}^3 & -A_{12}^3 & -A_{13}^3 & A_{14}^i s_i - A_{14}^3 \\ -A_{21}^3 & I_{k_1} s_1 - A_{22}^3 & -A_{23}^3 & -A_{24}^3 \\ -A_{31}^3 & -A_{32}^3 & I_{k_2} s_2 - A_{33}^3 & -A_{34}^3 \\ \hline A_{41}^\iota s_\iota - A_{41}^3 & -A_{42}^3 & -A_{43}^3 & s_3 - A_{44}^3 \end{bmatrix} \tag{4.45}$$

where $i = 1,2; \iota = \begin{cases} 1 & i = 2 \\ 2 & i = 1 \end{cases}$. The system matrix (4.45) can clearly be considered as

the representation of the following general state-space model of 2.D systems, which lies between the Fornasini-Marchesini and Roesser models (i=1)

$$\kappa(i_1 + 1, i_2 + 1) = A_{11}^1 \kappa(i_1 + 1, i_2) + A_{11}^2 \kappa(i_1, i_2 + 1) + A_{11}^3 \kappa(i_1, i_2) + A_{12}^3 \kappa^h(i_1, i_2)$$

$$+ A_{13}^3 \kappa^v(i_1, i_2) + A_{14}^3 u(i_1, i_2) + A_{14}^1 u(i_1 + 1, i_2),$$

$$\kappa^h(i_1 + 1, i_2) = A_{21}^3 \kappa(i_1, i_2) + A_{22}^3 \kappa^h(i_1, i_2) + A_{23}^3 \kappa^v(i_1, i_2) + A_{24}^3 u(i_1, i_2),$$

$$\kappa^v(i_1, i_2 + 1) = A_{31}^3 \kappa(i_1, i_2) + A_{32}^3 \kappa^h(i_1, i_2) + A_{33}^3 \kappa^v(i_1, i_2) + A_{34}^3 u(i_1, i_2), \tag{4.46}$$

$$y(i_1, i_2) = A_{41}^2 \kappa(i_1, i_2 + 1) + A_{41}^3 \kappa(i_1, i_2) + A_{42}^3 \kappa^h(i_1, i_2)$$

$$(4.47)$$

$$+ A_{43}^3 \kappa^v(i_1, i_2) + A_{44}^3 u(i_1, i_2).$$

The partial state vector here is the triple of sub-vectors

$$x(i_1, i_2) = \begin{bmatrix} \kappa(i_1, i_2) \\ \kappa^h(i_1, i_2) \\ \kappa^v(i_1, i_2) \end{bmatrix}$$

$$(4.48)$$

and resembles those from both the Fornasini-Marchesini and Roesser models. Note, however, that in the general case, as in Theorem 4.3, there is updating in the output equation of (4.47), due to the state sub-vector terms κ^h or κ^v, and this clearly leads to a new version of model non-causality. If we wish to obtain an output equation without a time shift we clearly require that $A_{41}^1 = 0$ or $A_{41}^2 = 0$.

Even with these singularities, the system matrix (4.45)can be transformed to the Roesser form by application of the EOA. The result is given below for the case of $i=1$. The case of $i=2$, is virtually identical and is hence omitted here.

Theorem 4.5 *The system matrix (4.45) for i=1 can be transformed by using the EOA to the following nonsingular Roesser form*

$$\begin{bmatrix} I_n s_1 - A_{11}^2 & -A_{12}^3 & A_{11}^3 + A_{11}^2 A_{11}^1 & -A_{13}^3 & -A_{14}^3 + A_{11}^2 A_{14}^1 \\ 0 & I_{k_1} s_1 - A_{22}^3 & A_{21}^3 & -A_{23}^3 & -A_{24}^3 \\ I_n & 0 & I_n s_2 - A_{11}^1 & 0 & -A_{14}^1 \\ 0 & -A_{32}^3 & A_{31}^3 & I_{k_2} s_2 - A_{33}^3 & -A_{34}^3 \\ \hline A_{41}^2 & -A_{42}^3 & A_{41}^3 - A_{41}^2 A_{11}^1 & -A_{43}^3 & s_3 - A_{44}^3 - A_{41}^2 A_{14}^1 \end{bmatrix}. \quad (4.49)$$

Proof: Apply the chain of operations

$$\mathbf{BlockAugment}_n(\Theta(s_1, s_2)), \quad \mathbf{L}(2+1\times(I_n s_1 - A_{11}^2)), \; \mathbf{L}(5+1\times(A_{41}^2)),$$

$$\mathbf{R}(2+1\times(-I_n s_2 + A_{11}^1)), \; \mathbf{R}(5+1\times(-A_{14}^1)) \qquad (4.50)$$

and follow this by appropriate row and column permutations and multiplications by"-1" to yield the required form of (4.49). ∎

Hence, the following standard state-space realization of the Roesser type has been obtained:

$$A = \begin{bmatrix} A_{11}^2 & A_{12}^3 & -A_{11}^3 - A_{11}^2 A_{11}^1 & A_{13}^3 \\ 0 & A_{22}^3 & -A_{21}^3 & A_{23}^3 \\ -I_n & 0 & A_{11}^1 & 0 \\ 0 & A_{32}^3 & -A_{31}^3 & A_{33}^3 \end{bmatrix}, B = \begin{bmatrix} A_{14}^3 - A_{11}^2 A_{14}^1 \\ A_{24}^3 \\ A_{14}^1 \\ A_{34}^3 \end{bmatrix}$$

$$\tag{4.51}$$

$$C = \begin{bmatrix} -A_{41}^2 & A_{42}^3 & -A_{41}^3 + A_{41}^2 A_{11}^1 & A_{43}^3 \end{bmatrix} D = A_{44}^3 + A_{41}^2 A_{14}^1.$$

5. MIMO Systems – the 1D Case

The key objective of this chapter is to generalize the previous results for SISO systems to the MIMO case. In this chapter we consider the 1D case as necessary preliminary analysis for the 2D/nD cases, which follow in the next chapter. Hence, the starting point is the univariate rational matrix

$$F(s) = \left[f_{ij}(s) \right]_{i=1,2,\dots,q;\, j=1,2,\dots,p} \tag{5.1}$$

where

$$f_{ij}(s) = \frac{b_{ij}(s)}{a_{ij}(s)} . \tag{5.2}$$

and $a_{ij}(s)$, $b_{ij}(s)$ are polynomials in s.

To set up the most obvious approach to generalization of the previous results, replace all rational elements $f_{ij}(s)$ of the matrix (5.1) by its bivariate polynomial counterparts as has been demonstrated in the previous chapters. Then the result is the bivariate polynomial matrix

$$A_f(s,z) = z\hat{A}(s) - \hat{B}(s), \tag{5.3}$$

where

$$\hat{A}(s) \triangleq \left[a_{ij}(s) \right],\ \hat{B}(s) \triangleq \left[b_{ij}(s) \right]. \tag{5.4}$$

Also, it follows immediately that applying the EOA to the matrix (5.3), considered as a collection of its elements, yields the following matrix of the required form

$$\left[\begin{array}{c|c} sI_N - \tilde{A} & -\tilde{B} \\ \hline -\tilde{C} & z\mathbf{1}_{q,p} - \tilde{\tilde{D}} \end{array} \right], \tag{5.5}$$

where

$$\mathbf{1}_{q,p} \triangleq \begin{bmatrix} 1 & \cdots & 1 \\ & \cdots & \\ 1 & \cdots & 1 \end{bmatrix}_{q \times p} . \tag{5.6}$$

The operator Augment can easily be generalized for rectangular matrices, *i.e.*

$$\text{Augment}: \text{Mat}_{q,p}(\Re[s,z]) \to \text{Mat}_{q+1,p+1}(\Re[s,z]), \tag{5.7}$$

where $\Re[s,z]$ denotes the ring of polynomials in s and z, and is defined in the same way as (3.2). Next, it is clear that the matrix (5.5) can be transformed as follows:

$$\left[\begin{array}{c|c} sI_N - \tilde{A} & -\tilde{B} \\ \hline -\tilde{C} & z1_{q,p} - \tilde{D} \end{array}\right] = \left[\begin{array}{c|c} sI_N - \tilde{A} & 0 \\ \hline 0 & I_p \end{array}\right]\left[\begin{array}{c|c} I_N & 0 \\ \hline -\tilde{C} & I_p \end{array}\right]$$

$$\times \left[\begin{array}{c|c} I_N & 0 \\ \hline 0 & z1_{q,p} - \tilde{D} - \tilde{C}(sI_t - \tilde{A})^{-1}\tilde{B} \end{array}\right]\left[\begin{array}{c|c} I_N & -(sI_N - \tilde{A})^{-1}\tilde{B} \\ \hline 0 & I_p \end{array}\right], \tag{5.8}$$

where the matrix $z1_{q,p} - \tilde{D} - \tilde{C}(sI_N - \tilde{A})^{-1}\tilde{B}$ represents $z1_{q,p} - \tilde{F}(s)$ and $\tilde{F}(s)$ is some transfer function matrix. The key problem is now that, in general,

$$\tilde{F}(s) \neq F(s). \tag{5.9}$$

The reason for this situation is that we are not allowed to apply some elementary operations, which destroy the transfer function matrix structure. A general solution to this key problem is developed below, but first the following illustrative example is presented.

Example 5.1 Consider the transfer function matrix

$$F(s) = \left[\begin{array}{cc} \dfrac{s+1}{s^2+1} & \dfrac{2}{s^2-2} \end{array}\right].$$

Then, the matrix $A_f(s,z)$ is defined as

$$A_f(s,z) = \left[z(s^2+1)-s-1 \quad z(s^2-2)-2\right].$$

Apply the following procedure:

$\quad$ Augment($A_f(s,z)$), L(2+1×z), R(2+1×(−s^2−1)), R(3+1×(−s^2+2)) → A_{f1},

$\quad$ Augment(A_{f1}), L(2+1×s), L(2+1×1), R(3+1×s) → A_{f2},

$\quad$ Augment(A_{f2}), L(3+1×s), R(2+1×(−1)), R(5+1×s), R(2+3×1) → A_{f3}

and follow this by applying the suitable row and column permutations to A_{f3} to yield

$$\left[\begin{array}{c|c} sI_3 - \tilde{A} & -\tilde{B} \\ \hline -\tilde{C} & z1_{1,2} - \tilde{D} \end{array}\right] = \left[\begin{array}{ccc|cc} s & 0 & 1 & -1 & 0 \\ 0 & s & 0 & 1 & 0 \\ 2 & -1 & s & 1 & 1 \\ \hline -2 & -1 & 0 & z+1 & z \end{array}\right].$$

Note, however, that

$$\tilde{D} + \tilde{C}\left(sI_t - \tilde{A}\right)^{-1}\tilde{B} = \left[-1 + \dfrac{s^2 + 2s + 4}{s(s^2 - 2)} \quad \dfrac{2s}{s^2 - 2}\right],$$

i.e. $\tilde{F}(s) \neq F(s)$.

The key to removing these difficulties is to assign more than one additional variable to the elements of the transfer function matrix. In effect, these additional variables act as 'frames' for the application of elementary operations in a manner, which leaves the transfer function invariant. Before deriving the general solution, it is instructive to consider some special cases, which will illustrate the main steps in establishing the general case.

5.1 Multi-input Single-output Systems

Here, row rational matrices are considered (as the simplest case) and it is shown that in such a case the EOA yields proper results. Instead of the representation (5.3), we use the multivariate representation

$$A_f(s, z_1, z_2, \ldots, z_p) = \left[\cdots \quad z_j a_j(s) - b_j(s) \quad \cdots\right]_{j=1,2,\ldots,p}. \qquad (5.10)$$

where now the transfer function matrix is the row matrix

$$F(s) = \left[\cdots \quad \dfrac{b_j(s)}{a_j(s)} \quad \cdots\right]_{j=1,2,\ldots,p}. \qquad (5.11)$$

In this case it is easily seen that each element possesses the necessary additional variable and from this it follows that the EOA works well here. The general procedure for this can be stated as follows:

First, apply the following operations to the polynomial matrix (5.10):

Augment($A_f(s, z_1, \cdots, z_p)$), L(2+1×z_1), R(2+1×($-a_1(s)$)) $\rightarrow A_{f1}$,

Augment(A_{f1}), L(3+1×z_2), R(4+1×($-a_2(s)$)) $\rightarrow A_{f2}$,

$\cdots$

Augment($A_{f,p-1}$), L((p+1)+1×z_p), R(2p+1×($-a_p(s)$)) $\rightarrow A_{fp}$,   (5.12)

to yield the matrix

$$
\begin{bmatrix}
1 & \cdots & 0 & 0 & 0 & 0 & \cdots & -a_p(s) \\
\vdots & \ddots & \vdots & \vdots & \vdots & \vdots & \ddots & \vdots \\
0 & \cdots & 1 & 0 & 0 & -a_2(s) & \cdots & 0 \\
0 & \cdots & 0 & 1 & -a_1(s) & 0 & \cdots & 0 \\
z_p & \cdots & z_2 & z_1 & -b_1(s) & -b_2(s) & \cdots & -b_p(s)
\end{bmatrix}.
\tag{5.13}
$$

Next, we apply a procedure similar to those presented previously for the SISO case. In particular, note that each pair of polynomials $a_i(s)$ and $b_i(s)$ $i=1,2,\ldots,p$ can be written in the form

$$
a_i(s) = (s+\alpha_1)a_i^1(s) + x_1^1, b_i(s) = (s+\alpha_1)b_i^1(s) + x_2^1,
\tag{5.14}
$$

where $x_i^1, \alpha_i \in \mathfrak{R}, i=1,2$ and the polynomials $a_i^1(s)$ and $b_i^1(s)$ are of the degree one lower than $a_i(s)$ and $b_i(s)$ respectively. Hence, apply the following operations (only the transformed column is considered):

$$
\text{Augment}\begin{bmatrix} -a_i(s) \\ -b_i(s) \end{bmatrix}, \text{L}(2+1\times(a_i^1(s)), \text{L}(3+1\times(b_i^1(s)),
$$

$$
\text{R}(2+1\times(s+\alpha_1))\rightarrow
\begin{bmatrix}
1 & s+\alpha_1 \\
a_i^1(s) & -x_1^1 \\
b_i^1(s) & -x_2^1
\end{bmatrix}.
\tag{5.15}
$$

Repeating these operations the required number of times for each $i=1,2,\ldots,p$ yields, after suitable row and column permutations, the required system matrix

$$
\begin{bmatrix}
sI_2 - A & \vdots & -B & \\
\hdashline
-C & \vdots & z_1 \quad z_2 \quad \cdots \quad z_p & -D
\end{bmatrix}.
\tag{5.16}
$$

Note that in this case is impossible to apply the previously discussed illegal operations. The following example provides a further illustration to this case.

Example 5.2 Consider the same transfer function as in Example 5.1 and introduce

$$
A_f(s, z_1, z_2) = \begin{bmatrix} z_1(s^2+1) - s - 1 & z_2(s^2-2) - 2 \end{bmatrix}.
$$

Now, apply the following procedure:

Augment($A_f(s, z_1, z_2)$), L(2+1×z_1), R(2+1×(−s²−1)) → A_{f1},

Augment(A_{f1}), L(2+1×s), L(3+1×1), R(3+1×s) → A_{f2},

Augment(A_{f2}), L(4+1×z_2), R(4+1×($-s^2$+2))) → A_{f3},

Augment(A_{f3}), L(2+1×s), R(6+1×s).

Note that, as illustrated here, the ordering of the steps in the general procedure can be changed as appropriate to the particular example under consideration. Finally, applying suitable row and column permutations yields the system matrix

$$H = \left[\begin{array}{cccc|cc} s & 1 & 0 & 0 & 0 & 0 \\ 2 & s & 0 & 0 & 0 & 1 \\ 0 & 0 & s & 1 & 0 & 0 \\ 0 & 0 & -1 & s & 1 & 0 \\ \hline -2 & 0 & -1 & 1 & z_1 & z_2 \end{array}\right]$$

and it is easy to check that this is the system matrix for the correct transfer function matrix in Example 5.1.

Unfortunately, this approach cannot be directly applied to multi-row polynomial matrices, which are the next special case discussed.

5.2 Single-input Multi-output Systems

As shown below, two approaches can be developed for this case, which differ in the number of additional variables employed. In both the cases the starting point is the column matrix

$$F(s) = \left[\begin{array}{c} \vdots \\ \dfrac{b_i(s)}{a_i(s)} \\ \vdots \end{array}\right]_{i=1,2,\ldots,q} \tag{5.17}$$

In the so-called first redundant approach, the first step is to define the polynomial matrix

$$A_f(s, z_1, z_2, \ldots, z_q) = \left[\begin{array}{c} \vdots \\ z_i a_i(s) - b_i(s) \\ \vdots \end{array}\right]_{i=1,2,\ldots,q} \tag{5.18}$$

Then separate at one step the variable s from the remaining ones. This, in turn, means that the matrix

$$
\begin{bmatrix}
1 & \beta_0(s) \\
z_1 - \alpha_1 & \beta_1(s) \\
z_2 - \alpha_2 & \beta_2(s) \\
\vdots & \vdots \\
z_q - \alpha_q & \beta_q(s)
\end{bmatrix},
$$

where $\beta_i(s)$, $i = 1,2,...,q$ are polynomials in s and α_i are real numbers, is required as a result of the first algorithm step. Generally, this last requirement is impossible when applying only elementary operations unless all denominators in the column matrix (5.17) are the same.

Instead, use the following procedure. First, reduce all polynomial fractions in the column matrix to a common denominator, *i.e.* write the transfer function matrix in the form

$$
F(s) = \begin{bmatrix} \vdots \\ \dfrac{B_i(s)}{A(s)} \\ \vdots \end{bmatrix}_{i=1,2,...,q}, \tag{5.19}
$$

where $A(s)$ is the least common multiple of all the denominators $a_i(s)$. Next, instead of (5.18), determine the polynomial matrix

$$
A_f(s, z_1, z_2, ..., z_q) = \begin{bmatrix} \vdots \\ z_i A(s) - B_i(s) \\ \vdots \end{bmatrix}_{i=1,2,...,q}. \tag{5.20}
$$

Finally, apply the following EOA operations:

Augment($A_f(s, z_1, ..., z_q)$), L(2+1×z_1), L(3+1×z_2), ...,L((q+1)+1×z_q),

R(2+1× $A(s)$) $\qquad$ (5.21)

which yield

$$
A_{f1} = \begin{bmatrix}
1 & -A(s) \\
z_1 & -B_1(s) \\
z_2 & -B_2(s) \\
\vdots & \vdots \\
z_q & -B_q(s)
\end{bmatrix}. \tag{5.22}
$$

Next, we apply a procedure similar to that expressed by (5.12). First, write the polynomials in (5.22) as

$$
A(s) = (s + \alpha_1)A^1(s) + x_0^1, \quad B_i(s) = (s + \alpha_1)B_i^1(s) + x_i^1; \quad i = 1,2,...,q, \tag{5.23}
$$

where α_1, $x_i^1 \in \Re, i = 0,1,\ldots,q$ and the polynomials $A^1(s)$ and $B_i^1(s)$ have the degrees one lower than $A(s)$ and $B_i(s)$, respectively. Then apply the operations

Augment(A_{f1}), L($2+1\times A^1(s)$), L($3+1\times B_1^1(s)$), ... ,L($(q+2)+1\times B_q^1(s)$),

$$R(3+1\times(s+\alpha_1)) \tag{5.24}$$

to obtain

$$A_{f2} = \begin{bmatrix} 1 & 0 & s+\alpha_1 \\ A^1(s) & 1 & -x_0^1 \\ B_1^1(s) & z_1 & -x_1^1 \\ B_2^1(s) & z_2 & -x_2^1 \\ \vdots & \vdots & \vdots \\ B_q^1(s) & z_q & -x_q^1 \end{bmatrix}. \tag{5.25}$$

Next, apply the same procedure to the column

$$\begin{bmatrix} A^1(s) \\ B_1^1(s) \\ B_2^1(s) \\ \vdots \\ B_q^1(s) \end{bmatrix} \tag{5.26}$$

i.e. set

$$A^1(s) = (s+\alpha_2)A^2(s) + x_0^2, \ B_i^1(s) = (s+\alpha_2)B_i^2(s) + x_i^2; \ i = 1,2,\ldots,q, \tag{5.27}$$

where α_2, $x_i^2 \in \Re, i = 0,1,\ldots,q$ and the polynomials $A^2(s)$ and $B_i^2(s)$ have the degrees one lower than $A^1(s)$ and $B_i^1(s)$, respectively. Now, apply the operations

Augment(A_{f2}), L($3+1\times(-A^2(s))$), L($4+1\times(-B_1^2(s))$),

$$L((q+3)+1\times(-B_q^2(s))), \ R(2+1\times(s+\alpha_2)) \tag{5.28}$$

to yield

$$
A_{f3} = \begin{bmatrix}
1 & s+\alpha_2 & 0 & 0 \\
0 & 1 & 0 & s+\alpha_1 \\
-A^2(s) & x_0^2 & 1 & -x_0^1 \\
-B_1^2(s) & x_1^2 & z_1 & -x_1^1 \\
-B_2^2(s) & x_2^2 & z_2 & -x_2^1 \\
\vdots & \vdots & \vdots & \vdots \\
-B_q^2(s) & x_q^2 & z_q & -x_q^1
\end{bmatrix}
\tag{5.29}
$$

and so on. This eventually results in a matrix of the form

$$
A_{f_t}(s,z_1,z_2,\ldots,z_q)=\begin{bmatrix}
1 & s+\alpha_{t-1} & \cdots & 0 & 0 & 0 \\
0 & 1 & \ddots & 0 & 0 & 0 \\
\vdots & \vdots & \ddots & \vdots & \vdots & \vdots \\
0 & 0 & \cdots & s+\alpha_2 & 0 & 0 \\
0 & 0 & \cdots & 1 & 0 & s+\alpha_1 \\
\varepsilon_0 s+\phi_0 & (+/-)x_0^{t-1} & \cdots & x_0^2 & 1 & -x_0^1 \\
\varepsilon_1 s+\phi_1 & (+/-)x_1^{t-1} & \cdots & x_1^2 & z_1 & -x_1^1 \\
\varepsilon_2 s+\phi_2 & (+/-)x_2^{t-1} & \cdots & x_2^2 & z_2 & -x_2^1 \\
\vdots & \vdots & \cdots & \vdots & \vdots & \vdots \\
\varepsilon_q s+\phi_q & (+/-)x_q^{t-1} & \cdots & x_q^2 & z_q & -x_q^1
\end{bmatrix},
\tag{5.30}
$$

where t denotes now the degree of $A(s)$. If a common denominator $A(s)$ of $a_i(s),\ i=1,2,\ldots,q$ has a degree greater than each $B_i(s),\ i=1,2,\ldots,q$, then all $\varepsilon_i = 0$. If, however, there is at least one $B_i(s), i=1,2,\ldots,q$ with degree equal to that of $A(s)$, then the following additional operations must be applied:

$$
L\left((t+k)+t\times\left(-\frac{\varepsilon_k}{\varepsilon_0}\right)\right),\ k\in 1,2,\ldots,q.
\tag{5.31}
$$

Finally, employing suitable row and column permutations yields the required system matrix

$$
\begin{bmatrix}
sI_t - A & -B \\
\hline
-C & \begin{matrix} z_1 \\ z_2 \\ \vdots \\ z_q \end{matrix} \ -D
\end{bmatrix}.
\tag{5.32}
$$

In this case, the sequence of operations applied is defined uniquely, but there are numerous possibilities for changing their order of application to obtain various appropriate results. Note now that the use of certain non-augmenting elementary row operations within the block

$$\begin{bmatrix} -B_1(s) \\ -B_2(s) \\ \vdots \\ -B_q(s) \end{bmatrix}$$

would change the transfer function matrix. However, these operations are forbidden due to the form of the block

$$\begin{bmatrix} z_1 \\ z_2 \\ \vdots \\ z_q \end{bmatrix}.$$

Row elementary operations on this matrix simply yield some combinations of additional, artificial variables z_i, $i=1,2,...,q$, which inform us of illegal operations. The following example illustrates this analysis.

Example 5.3. Consider the matrix

$$F(s) = \begin{bmatrix} \dfrac{1}{s+1} \\ \dfrac{-1}{s-1} \end{bmatrix}.$$

Then the matrix $A_f(s, z_1, z_2)$ takes the form

$$A_f(s, z_1, z_2) = \begin{bmatrix} z_1(s^2-1) - s + 1 \\ z_2(s^2-1) + s + 1 \end{bmatrix}.$$

Now, apply the operations:

Augment($A_f(s, z_1, z_2)$), L(2+1×z$_1$), L(3+1×z$_2$), R(2+1×(−s²+1)) → A_{f1},

Augment(A_{f1}), L(2+1×s), L(3+1×1), L(4+1×(−1)), R(3+1×s) → A_{f2}.

Finally, applying appropriate row and column permutations yields the system matrix

$$\left[\begin{array}{c|c} sI_2 - A & -B \\ \hline -C & \begin{bmatrix} z_1 \\ z_2 \end{bmatrix} - D \end{array} \right] = \left[\begin{array}{cc|c} s & 1 & 0 \\ 1 & s & 1 \\ \hline 1 & 1 & z_1 \\ 1 & -1 & z_2 \end{array} \right].$$

It is easy to see that the matrix $D + C(sI_n - A)^{-1}B$ is equal to the given transfer function matrix.

The second, so-called restricted number of additional variables approach, is based on the fact that the number of additional variables used in the first approach can be restricted to less than $n+1$. First, return to (5.17) and note that, in principle, for this column matrix only one additional variable is required. Hence instead of the $q+1$-variate matrix (5.22), start with the bivariate matrix

$$
A_f(s,z) = \begin{bmatrix} zA(s) - B_1(s) \\ -B_2(s) \\ \vdots \\ -B_q(s) \end{bmatrix}
\tag{5.33}
$$

and apply the algorithm. In particular, apply

$$
\text{Augment}(\, A_f(s,z)\,),\ \text{L}(2+1\times z),\ \text{R}(2+1\times A(s))
\tag{5.34}
$$

to obtain

$$
A_{f1} = \begin{bmatrix} 1 & -A(s) \\ z & -B_1(s) \\ 0 & -B_2(s) \\ \vdots & \vdots \\ 0 & -B_q(s) \end{bmatrix}.
\tag{5.35}
$$

Then apply the same procedure used to obtain (5.17) to yield the required system matrix

$$
\left[\begin{array}{c|c} sI_t - A & -B \\ \hline & \begin{array}{c} z \\ 0 \\ \vdots \\ 0 \end{array} \\ -C & -D \end{array} \right].
\tag{5.36}
$$

Remark In this form of the algorithm, it is essential not to apply at each step any non-augmenting row elementary operations within the columns

$$
\mathbf{Z} \triangleq \begin{bmatrix} z \\ 0 \\ \vdots \\ 0 \end{bmatrix}
$$

and

$$\begin{bmatrix} -B_1(s) \\ -B_2(s) \\ \vdots \\ -B_q(s) \end{bmatrix}$$

since they change the final transfer function matrix $F(s) = D + C(sI_t - A)^{-1}B$ and the form of Z does not prevent this change.

Example 5.4. Consider again the transfer function matrix of Example 5.3 and introduce the polynomial matrix

$$A_f(s,z) = \begin{bmatrix} z(s^2 - 1) - s + 1 \\ s + 1 \end{bmatrix}.$$

Then, apply the operations:

Augment($A_f(s, z_1, z_2)$), L(2+1×z_1), R(2+1×($-s^2+1$)) → A_{f1},

Augment(A_{f1}), L(2+1×s), L(3+1×1), L(4+1×(-1)), R(3+1×s) → A_{f2}

and continue as required. Here, as compared to the procedure used in Example 5.3, the operation L(3+1×z_2) has been removed but the same result has been obtained.

5.3 MISO and SIMO Cases – the Dual Approach

Consider first the SIMO case, *i.e.* the column transfer matrix and look for the possibility of avoiding the representation (5.19) with the common denominator and proceed along the similar lines as for the MISO case (the row transfer function matrix). The first immediate approach is to transpose the column transfer function matrix, and then proceed as per the row matrix case and finally transpose. Also, one can produce the equivalent approach to that used for the row transfer function matrix. Start, hence, from (5.18) and instead of the procedure shown in the previous sub-section apply:

Augment($A_f(s, z_1, \cdots, z_p)$), R(2+1×z_1), L(2+1×($-a_1(s)$)) → A_{f1},

Augment(A_{f1}), R(3+1×z_2), L(4+1×($-a_2(s)$))) → A_{f2},

...   (5.37)

Augment($A_{f,p-1}$), R((p+1)+1×z_p), L($2p$+1×($-a_p(s)$))) → A_{fp},

to yield the matrix

$$
\begin{bmatrix}
1 & \cdots & 0 & 0 & z_p \\
\vdots & \ddots & \vdots & \vdots & \vdots \\
0 & \cdots & 1 & 0 & z_2 \\
0 & \cdots & 0 & 1 & z_1 \\
0 & \cdots & & -a_1(s) & -b_1(s) \\
 & & -a_2(s) & & -b_2(s) \\
 & \cdot{}^{\cdot} & & \vdots & \vdots \\
-a_p(s) & \cdots & & 0 & -b_p(s)
\end{bmatrix}.
\tag{5.38}
$$

The next steps of the algorithm are standard and hence are omitted here.

Example 5.5 Consider the transfer function

$$
F(s) = \begin{bmatrix} \dfrac{s-1}{s+1} \\ \dfrac{1}{s+2} \end{bmatrix}.
$$

It is straightforward to write the representation (5.38) as

$$
\begin{bmatrix}
1 & 0 & z_2 \\
0 & 1 & z_1 \\
0 & -s-1 & -s+1 \\
-s-2 & 0 & -1
\end{bmatrix}.
$$

Here, it is sufficient to multiply the second column by '-1' and add the result to the third. Final respective row multiplications by '-1' and row and column permutations yield the system matrix

$$
\begin{bmatrix}
s+2 & 0 & 1 \\
0 & s+1 & -2 \\
0 & 1 & z_1-1 \\
1 & 0 & z_2
\end{bmatrix},
$$

which constitutes the following state-space realization

$$
A = \begin{bmatrix} -2 & 0 \\ 0 & -1 \end{bmatrix}, \quad
b = \begin{bmatrix} -1 \\ 2 \end{bmatrix}, \quad
C = \begin{bmatrix} 0 & -1 \\ -1 & 0 \end{bmatrix}, \quad
D = \begin{bmatrix} 1 \\ 0 \end{bmatrix}.
$$

It is easy to check that this state-space realization gives the required transfer function matrix.

Return now to the MISO case (row transfer function matrix). Then, direct application the above procedure is impossible. There is, however, a similar possibility as for SIMO systems, *i.e.* one can transform the transfer function row to the equivalent form with one common denominator. Hence, return to (5.10) and represent it as

$$A_f(s, z_1, z_2, \ldots, z_p) = \left[\cdots \quad z_j A(s) - B_j(s) \quad \cdots \right]_{j=1,2,\ldots,p} , \qquad (5.39)$$

where $A(s)$ is the least common denominator of all the rational functions $f_j(s)$ and $B_j(s)$ are the new numerators in the representation of $f_j(s)$, *i.e.*

$$f_j(s) = \frac{b_j(s)}{a_j(s)} = \frac{B_j(s)}{A(s)}$$

or equivalently, *c.f.* (5.33)

$$A_f(s, z) = \left[zA(s) - B_1(s) \quad -B_2(s) \quad \cdots \quad -B_p(s) \right] . \qquad (5.40)$$

Next apply as for the case of (5.40):

Augment($A_f(s, z)$), R(2+1×z), L(2+1×(− $A(s)$)) → A_{f1},

and subsequently proceed in much the same way as in (5.21)-(5.31) but in the horizontal direction. The procedure is virtually the same as before and hence we do not present the full algorithm. It is, however, illustrated by the following example.

Example 5.6 Consider the same transfer function as in Examples 1 and 2 and introduce instead of

$$A_f(s, z_1, z_2) = \left[z_1(s^2 + 1) - s - 1 \quad z_2(s^2 - 2) - 2 \right]$$

the matrix polynomial A_f of the form of (5.40), *i.e.*

$$A_f(s, z) = \left[z(s^2 + 1)(s^2 - 2) - (s + 1)(s^2 - 2) \mid -2(s^2 + 1) \right] .$$

Now, apply the following procedure:

```
> B1:=Augment(Af);
> B1:=addcol(B1,1,2,z);
> B1:=addrow(B1,1,2,-(s^2+1)*(s^2-2));
> B2:=Augment(B1);
> B2:=addrow(B2,1,3,s^2-2);
> B2:=addcol(B2,1,2,s^2+1);
> B2:=addcol(B2,1,3,s+1);
```

```
> B2:=addcol(B2,1,4,2);
```

```
> B3:=Augment(B2);
```

```
> B3:=addrow(B3,1,2,s);
```

```
> B3:=addcol(B3,1,3,-s);
```

```
> B3:=addcol(B3,1,4,-1);
```

```
> B4:=Augment(B3);
```

```
> B4:=addrow(B4,1,5,-s);
```

```
> B4:=addcol(B4,1,3,s);
```

```
> B4:=mulrow(B4,2,-1);
```

```
> B4:=mulrow(B4,5,-1);.
```

Finally, applying the following row and column permutations:

```
> pr:=[5,3,1,2,4]:      #permutation for rows
```

```
> pc:=[1,2,3,4,5,6]: #permutation for columns
```

yields the system matrix

$$
H = \left[\begin{array}{cccc:cc}
s & 0 & 2 & 0 & 0 & 6 \\
0 & s & 1 & 1 & 1 & 2 \\
1 & 0 & s & 0 & 0 & 0 \\
0 & -1 & 0 & s & 1 & 0 \\
\hdashline
0 & 0 & 0 & 1 & z & 0
\end{array}\right]
$$

and it is easy to check that this is the system matrix for the correct transfer function matrix in Example 5.1. However, the dimension of the realization increased by one.

Example 5.7 Here we use virtually the same initial representation as the previous example and use the full number of additional variables. This is because using a restricted number of such variables can lead to some difficulty in partitioning the resulting system matrix into the form necessary to obtain the state-space realization matrices – and, at worst, a singular realization could result. Consider, therefore, the matrix

$$
A_f(s, z_1, z_2)
$$
$$
= \left[z_1(s^2+1)(s^2-2) - (s+1)(s^2-2) \;\vdots\; z_2(s^2+1)(s^2-2) - 2(s^2+1) \right]
$$

and apply the following operations:

```
> B1:=Augment(A_f);
```

```
> B1:=addrow(B1,1,2,z1);
```

```
> B1:=addcol(B1,1,2,-(s^2+1)*(s^2-2));

> B2:=Augment(B1);

> B2:=addrow(B2,1,3,z2);

> B2:=addcol(B2,1,4,-(s^2+1)*(s^2-2));

> B3:=Augment(B2);

> B3:=addcol(B3,1,4,s^2-2);

> B3:=addrow(B3,1,4,s+1);

> B3:=addrow(B3,1,3,s^2+1);

> B4:=Augment(B3);

> B4:=addcol(B4,1,6,s^2+1);

> B4:=addrow(B4,1,3,s^2-2);

> B4:=addrow(B4,1,5,2);

> B5:=Augment(B4);

> B5:=addrow(B5,1,2,-s);

> B5:=addcol(B5,1,7,s);

> B6:=Augment(B5);

> B6:=addcol(B6,1,7,s);

> B6:=addrow(B6,1,4,-s);

> B7:=Augment(B6);

> B7:=addcol(B7,1,4,s);

> B7:=addrow(B7,1,6,-s);

> B8:=Augment(B7);

> B8:=addcol(B8,1,6,s);

> B8:=addrow(B8,1,8,-s);

> B8:=addrow(B8,1,9,-1);

> B8:=mulrow(B8,5,-1);

> B8:=mulrow(B8,6,-1);

> B8:=mulrow(B8,7,-1);

  B8:=mulrow(B8,8,-1);.
```

Finally, apply the following row and column permutations:

```
> pr:=[8,7,6,5,2,1,3,4,9]:        #permutation for rows

> pc:=[1,2,3,4,5,6,9,10,8,7]: #permutation for columns
```

to yield

$$
\begin{bmatrix}
s & 0 & 0 & 0 & 0 & -1 & 0 & 0 & -1 & 0 \\
0 & s & 0 & 0 & 2 & 0 & 0 & 0 & 0 & -1 \\
0 & 0 & s & 0 & 0 & -1 & 2 & 0 & 0 & 0 \\
0 & 0 & 0 & s & -1 & 0 & 0 & -1 & 0 & 0 \\
0 & 1 & 0 & 0 & s & 0 & 0 & 0 & 0 & 0 \\
1 & 0 & 0 & 0 & 0 & s & 0 & 0 & 0 & 0 \\
0 & 0 & 1 & 0 & 0 & 0 & s & 0 & 0 & 0 \\
0 & 0 & 0 & 1 & 0 & 0 & 0 & s & 0 & 0 \\
-1 & 0 & 0 & 0 & 2 & 1 & 0 & 0 & z_1 & z_2
\end{bmatrix},
$$

which gives a correct but highly redundant (non-minimal) solution. This strongly suggests that the EOA should be carefully applied to prevent the highly unsatisfactory result of solutions with very large dimensions.

5.4 General MIMO Case

First, note that for this case the techniques developed above for the MISO and SIMO cases which do not require the computation of the least common denominator of the matrix rows or columns is not possible. Consider, for notational simplicity, the following special case of the initial polynomial matrix

$$
\begin{bmatrix}
z_{11}a_{11} - b_{11} & z_{12}a_{12} - b_{12} \\
z_{21}a_{21} - b_{21} & z_{22}a_{22} - b_{22}
\end{bmatrix},
$$

where a_{ij} are polynomials. Now, applying a natural chain of the EOA operations for the MISO case yields the following polynomial matrix as an alternative to (5.13):

$$
\begin{bmatrix}
1 & 0 & 0 & 0 & 0 & -a_{22} \\
0 & 1 & 0 & 0 & -a_{21} & 0 \\
0 & 0 & 1 & 0 & 0 & -a_{12} \\
0 & 0 & 0 & 1 & -a_{11} & 0 \\
0 & 0 & z_{12} & z_{11} & -b_{11} & -b_{12} \\
z_{22} & z_{21} & 0 & 0 & -b_{21} & -b_{22}
\end{bmatrix},
$$

which loses its links to the investigated transfer function matrix. A similar situation occurs when applying the technique of the SIMO systems, c.f. (5.37)-(5.38).

Now, we can use two basic approaches: the first of which uses the least common denominator for the column matrices, and the second uses the least common denominator for the row matrices. Start with the first one and represent the given transfer function matrix as

$$f_{ij}(s) = \frac{b_{ij}(s)}{a_{ij}(s)} = \frac{B_{ij}(s)}{A_j(s)}, \ i = 1,2,\ldots,q, j = 1,2,\ldots,p, \tag{5.41}$$

where $A_j(s)$ denotes a common denominator of all the elements in the jth column of this transfer function matrix, which is clearly equivalent to the polynomial matrix description

$$F(s) = N_r(s)D_r(s)^{-1} \tag{5.42}$$

with diagonal matrix $D_r(s)$. In order to extend the approach to this general case, introduce the multivariate polynomial matrix

$$A_f(s, z_{11}, z_{12}, \ldots, z_{1m} \ldots z_{nm}) = \begin{bmatrix} \vdots \\ \cdots & z_{ij} A_j(s) - B_{ij}(s) & \cdots \\ \vdots \end{bmatrix}_{i=1,2,\ldots,q; j=1,2\ldots p} . \tag{5.43}$$

The second approach is based on representing the given transfer function matrix as

$$f_{ij}(s) = \frac{b_{ij}(s)}{a_{ij}(s)} = \frac{B_{ij}(s)}{\tilde{A}_i(s)}, \ i = 1,2,\ldots,q, j = 1,2,\ldots,p, \tag{5.44}$$

where $\tilde{A}_i(s)$ denotes a common denominator of all the elements in the ith row of this transfer function matrix, which is clearly equivalent to the polynomial matrix description

$$F(s) = D_l(s)^{-1} N_l(s) \tag{5.45}$$

with diagonal matrix $D_l(s)$. The matrix of (5.43) is now

$$A_f(s, z_{11}, z_{12}, \ldots, z_{1p} \ldots z_{qp}) = \begin{bmatrix} \vdots \\ \cdots & z_{ij} \tilde{A}_i(s) - B_{ij}(s) & \cdots \\ \vdots \end{bmatrix}_{i=1,2,\ldots,q; j=1,2\ldots p} \tag{5.46}$$

and it is clear that applying the EOA to these polynomial matrices yields the desired system matrix. Note, however, that this approach requires a large number of additional variables. In fact, the number of additional variables required can be significantly restricted as mentioned previously. To detail this generalisation replace (5.43) by the polynomial matrix

$A_f(s, z_1, \cdots, z_p)$

$$
= \begin{bmatrix}
z_1 A_1(s) - B_{11}(s) & z_2 A_2(s) - B_{12}(s) & & z_p A_p(s) - B_{1p}(s) \\
\quad - B_{21}(s) & \quad - B_{22}(s) & & \quad - B_{2p}(s) \\
\vdots & \vdots & \cdots & \vdots \\
\quad - B_{q1}(s) & \quad - B_{q2}(s) & & \quad - B_{qp}(s)
\end{bmatrix}. \tag{5.47}
$$

Then applying the first steps of the algorithm to each single column (c.f. Section 5.2) of this matrix yields the following matrix:

$$
\begin{bmatrix}
1 & \cdots & 0 & 0 & 0 & 0 & \cdots & - A_p(s) \\
\vdots & \ddots & \vdots & \vdots & \vdots & \vdots & \ddots & \vdots \\
0 & \cdots & 1 & 0 & 0 & - A_2(s) & \cdots & 0 \\
0 & \cdots & 0 & 1 & - A_1(s) & 0 & \cdots & 0 \\
z_p & \cdots & z_2 & z_1 & - B_{11}(s) & - B_{12}(s) & \cdots & - B_{1p}(s) \\
0 & \cdots & 0 & 0 & - B_{21}(s) & - B_{22}(s) & \cdots & - B_{2p}(s) \\
\cdots & \cdots & \cdots & \cdots & \cdots & \cdots & \cdots & \cdots \\
0 & \cdots & 0 & 0 & - B_{q1}(s) & - B_{q2}(s) & \cdots & - B_{qp}(s)
\end{bmatrix}. \tag{5.48}
$$

Next, applying the procedure to all the polynomial columns of the matrix (5.48) yields the required system matrix

$$
\begin{bmatrix}
sI_t - A & -B \\
\hline
-C & \begin{bmatrix} z_1 & z_2 & \cdots & z_p \\ 0 & 0 & \cdots & 0 \\ & \cdots & \cdots & \\ 0 & 0 & \cdots & 0 \end{bmatrix}_{q,p} \quad -D
\end{bmatrix}. \tag{5.49}
$$

Again, row elementary operations must not be used within the blocks

$$
\begin{bmatrix}
z_1 & z_2 & \cdots & z_p \\
0 & 0 & \cdots & 0 \\
& \cdots & \cdots & \\
0 & 0 & \cdots & 0
\end{bmatrix} \quad \text{and} \quad
\begin{bmatrix}
- B_{11}(s) & - B_{12}(s) & & - B_{1p}(s) \\
- B_{21}(s) & - B_{22}(s) & & - B_{2p}(s) \\
\vdots & \vdots & \cdots & \vdots \\
- B_{q1}(s) & - B_{q2}(s) & & - B_{qp}(s)
\end{bmatrix}
$$

at each step of the algorithm. Note that column operations within

$$
\begin{bmatrix}
z_1 & z_2 & \cdots & z_p \\
0 & 0 & \cdots & 0 \\
& \cdots & \cdots & \\
0 & 0 & \cdots & 0
\end{bmatrix}
$$

are not possible due the form of this matrix. The column operations within

$$\begin{bmatrix} -B_{11}(s) & -B_{12}(s) & & -B_{1p}(s) \\ -B_{21}(s) & -B_{22}(s) & \cdots & -B_{2p}(s) \\ \vdots & \vdots & & \vdots \\ -B_{q1}(s) & -B_{q2}(s) & & -B_{qp}(s) \end{bmatrix}$$

are allowed, which is established below.

Example 5.8 Consider the transfer function matrix

$$F(s) = \begin{bmatrix} \dfrac{1}{s+1} & \dfrac{1}{s} \\ \dfrac{1}{s-1} & \dfrac{1}{s+2} \end{bmatrix},$$

which can also be written as

$$F(s) = \begin{bmatrix} \dfrac{s-1}{s^2-1} & \dfrac{s+2}{s^2+2s} \\ \dfrac{s+1}{s^2-1} & \dfrac{s}{s^2+2s} \end{bmatrix}.$$

Now, define the matrix

$$A_f(s, z_1, z_2) = \begin{bmatrix} z_1(s^2-1)-s+1 & z_2(s^2+2s)-s-2 \\ -s-1 & -s \end{bmatrix}$$

and apply the following algorithm procedure:

Augment($A_f(s, z_1, z_2)$), L(2+1×z_1), R(2+1×($-s^2$+1)) → A_{f1},

Augment(A_{f1}), L(3+1×z_2), R(3+1×($-s^2$−2s)) → A_{f2},

Augment(A_{f2}), L(2+1×s), L(3+1×1), L(4+1×1), R(4+1×s) → A_{f3},

Augment(A_{f3}), L(3+1×(s+2)), L(5+1×1), L(6+1×1), R(6+1×s) → A_{f4}.

Finally, employing appropriate row and column permutations yields the required system matrix as

$$\begin{bmatrix} sI_4 - A & -B \\ \hline -C & \begin{bmatrix} z_1 & z_2 \\ 0 & 0 \end{bmatrix} - D \end{bmatrix} = \begin{bmatrix} s & 0 & 1 & 0 & 0 & 0 \\ 0 & s & 0 & 1 & 0 & 0 \\ 0 & 0 & s+2 & 0 & 0 & 1 \\ 0 & 1 & 0 & s & 1 & 0 \\ \hline -2 & 1 & 1 & 1 & z_1 & z_2 \\ 0 & -1 & 1 & 1 & 0 & 0 \end{bmatrix}$$

and it is easy to check that the matrix $D + C(sI_n - A)^{-1}B$ is equal to $F(s)$.

This example can also be used to highlight the errors arising from using illegal operations. In particular, start with A_{f2} and apply L(3+4×(−1)). Then apply the usual algorithm operations

$$\text{Augment}(A_{f2}), \; L(3+1\times s), \; L(5+1\times 1), \; R(4+1\times s) \; \rightarrow \; A_{f3},$$

$$\text{Augment}(A_{f3}), \; L(3+1\times(s+2)), \; L(6+1\times 1), \; R(6+1\times s) \; \rightarrow \; A_{f4}.$$

Appropriate row and column permutations yield now the system matrix

$$\begin{bmatrix} sI_4 - A & -B \\ \hline -C & \begin{bmatrix} z_1 & z_2 \\ 0 & 0 \end{bmatrix} - D \end{bmatrix} = \left[\begin{array}{cccc|cc} s & 0 & 1 & 0 & 0 & 0 \\ 0 & s & 0 & 1 & 0 & 0 \\ 0 & 0 & s+2 & 0 & 0 & 1 \\ 0 & 1 & 0 & s & 1 & 0 \\ \hline -2 & 2 & 0 & 0 & z_1 & z_2 \\ 0 & -1 & 1 & 1 & 0 & 0 \end{array}\right],$$

which leads to a wrong transfer function matrix, $i.e.$

$$D + C(sI_4 - A)^{-1}B = \begin{bmatrix} \dfrac{2}{1-s^2} & \dfrac{1}{s(s+2)} \\ \dfrac{1}{s-1} & \dfrac{1}{s+2} \end{bmatrix}.$$

Next, consider the 'dual case', $i.e.$ start by finding the least common denominators of the rows as opposed to the columns. In particular, start with an initial matrix of the form

$$A_f(s, z_1, \cdots, z_q) = \begin{bmatrix} z_1\tilde{A}_1(s) - B_{11}(s) & -B_{12}(s) & & -B_{1p}(s) \\ z_2\tilde{A}_2(s) - B_{21}(s) & -B_{22}(s) & \cdots & -B_{2p}(s) \\ \vdots & \vdots & & \vdots \\ z_q\tilde{A}_q(s) - B_{q1}(s) & -B_{q2}(s) & & -B_{qp}(s) \end{bmatrix}. \tag{5.50}$$

Then, applying the first steps of the algorithm to each single row of this matrix yields

$$\begin{bmatrix} 1 & \cdots & 0 & 0 & z_q & 0 & \cdots & 0 \\ \vdots & \ddots & \vdots & \vdots & \vdots & \vdots & \ddots & \vdots \\ 0 & \cdots & 1 & 0 & z_2 & 0 & \cdots & 0 \\ 0 & \cdots & 0 & 1 & z_1 & 0 & \cdots & 0 \\ 0 & \cdots & 0 & -\tilde{A}_1(s) & -B_{11}(s) & -B_{12}(s) & \cdots & -B_{1p}(s) \\ 0 & \cdots & -\tilde{A}_2(s) & 0 & -B_{21}(s) & -B_{22}(s) & \cdots & -B_{2p}(s) \\ \cdots & \cdots & \cdots & \cdots & \cdots & \cdots & \cdots & \cdots \\ -\tilde{A}_q(s) & \cdots & 0 & 0 & -B_{q1}(s) & -B_{q2}(s) & \cdots & -B_{qp}(s) \end{bmatrix}. \quad (5.51)$$

Next, applying the procedure to all the polynomial rows of the matrix (5.51) yields the required system matrix

$$\begin{bmatrix} sI_t - A & -B & \\ & \begin{bmatrix} z_1 & 0 & \cdots & 0 \\ z_2 & 0 & \cdots & 0 \\ & \cdots & \cdots & \\ z_q & 0 & \cdots & 0 \end{bmatrix}_{q,p} & \\ -C & & -D \end{bmatrix}. \quad (5.52)$$

Here, column elementary operations must not be used within the blocks

$$\begin{bmatrix} z_1 & 0 & \cdots & 0 \\ z_2 & 0 & \cdots & 0 \\ & \cdots & \cdots & \\ z_q & 0 & \cdots & 0 \end{bmatrix} \quad \text{and} \quad \begin{bmatrix} -B_{11}(s) & -B_{12}(s) & & -B_{1p}(s) \\ -B_{21}(s) & -B_{22}(s) & \cdots & -B_{2p}(s) \\ \vdots & \vdots & & \vdots \\ -B_{q1}(s) & -B_{q2}(s) & & -B_{qp}(s) \end{bmatrix}$$

at each step of the algorithm. Note that row operations within

$$\begin{bmatrix} z_1 & 0 & \cdots & 0 \\ z_2 & 0 & \cdots & 0 \\ & \cdots & \cdots & \\ z_q & 0 & \cdots & 0 \end{bmatrix}$$

are not possible due the form of this matrix. The row operations within

$$\begin{bmatrix} -B_{11}(s) & -B_{12}(s) & & -B_{1p}(s) \\ -B_{21}(s) & -B_{22}(s) & \cdots & -B_{2p}(s) \\ \vdots & \vdots & & \vdots \\ -B_{q1}(s) & -B_{q2}(s) & & -B_{qp}(s) \end{bmatrix}$$

are allowed, which is proved below.

Example 5.9 Consider again the transfer function matrix of Example 5.8, *i.e.*

$$F(s) = \begin{bmatrix} \dfrac{1}{s+1} & \dfrac{1}{s} \\ \dfrac{1}{s-1} & \dfrac{1}{s+2} \end{bmatrix},$$

which is now written as

$$F(s) = \begin{bmatrix} \dfrac{s}{(s+1)s} & \dfrac{s+1}{(s+1)s} \\ \dfrac{s+2}{(s-1)(s+2)} & \dfrac{s-1}{(s-1)(s+2)} \end{bmatrix}.$$

Next, define the matrix

$$A_f(s,z_1,z_2) = \begin{bmatrix} z_1(s+1)s - s & -s-1 \\ z_2(s-1)(s+2) - s - 2 & -s+1 \end{bmatrix}$$

and apply the following algorithm procedure:

```
> B1:=Augment(Af);
> B1:=addcol(B1,1,2,z1);
> B1:=addrow(B1,1,2,-(s+1)*s);
> B2:=Augment(B1);
> B2:=addcol(B2,1,3,z2);
> B2:=addrow(B2,1,4,-(s-1)*(s+2));
> B3:=Augment(B2);
> B3:=addrow(B3,1,4,s);
> B3:=addcol(B3,1,3,s+1);
> B3:=addcol(B3,1,4,1);
> B3:=addcol(B3,1,5,1);
> B4:=Augment(B3);
```

```
> B4:=addrow(B4,1,6,s-1);

> B4:=addcol(B4,1,3,s+2);

> B4:=addcol(B4,1,5,1);

> B4:=addcol(B4,1,6,1);.
```

Finally, apply the following row and column permutations:

```
> pr:=[6,5,1,2,4,3]:      #permutation for rows

> pc:=[1,2,3,4,5,6]:  #permutation for columns
```

to yield the alternative (to the previous example) system matrix:

$$\left[\begin{array}{cccc|cc} s-1 & 0 & 0 & 0 & -3 & 0 \\ 0 & s & 0 & 0 & 0 & -1 \\ 1 & 0 & s+2 & 0 & 1 & 1 \\ 0 & 1 & 0 & s+1 & 1 & 1 \\ \hline 0 & 0 & 0 & 1 & z_1 & 0 \\ 0 & 0 & 1 & 0 & z_2 & 0 \end{array}\right].$$

It is easy to check that this matrix gives the correct transfer function matrix for this example.

5.5 Further Analysis

In what follows, the cases $q \geq p$ (the number of rows greater than or equal to the number of columns) and $q < p$ (the number of rows less then a number of columns) will be separately investigated. Suggestions on the choice of the additional variable matrix with a restricted number of variables, equivalent to

$$\begin{bmatrix} z_1 & 0 & \cdots & 0 \\ z_2 & 0 & \cdots & 0 \\ & \cdots & \cdots & \\ z_q & 0 & \cdots & 0 \end{bmatrix} \quad \text{or} \quad \begin{bmatrix} z_1 & z_2 & \cdots & z_p \\ 0 & 0 & \cdots & 0 \\ & \cdots & \cdots & \\ 0 & 0 & \cdots & 0 \end{bmatrix},$$

are provided.

Case 1: $q \geq p$

In this case the number of additional variables in the matrix can be limited to a single one. Start with the initial representation

$$A_f(s,z) = \begin{bmatrix} zA_1(s) - B_{11}(s) & -B_{12}(s) & -B_{1p}(s) \\ -B_{21}(s) & zA_2(s) - B_{22}(s) & -B_{2p}(s) \\ \vdots & \vdots & \cdots & \vdots \\ -B_{p1}(s) & -B_{p2}(s) & zA_p(s) - B_{pp}(s) \\ \vdots & \vdots & \cdots & \vdots \\ -B_{q1}(s) & -B_{q2}(s) & -B_{qp}(s) \end{bmatrix}. \qquad (5.53)$$

and introduce the notation

$$\mathbf{B}(s) = \left[B_{ij}(s) \right]_{q,p} = \begin{bmatrix} \mathbf{B}_1(s) \\ \mathbf{B}_2(s) \end{bmatrix} \begin{matrix} \}p \\ \}q-p \end{matrix}. \qquad (5.54)$$

Now, apply the block operations

$$\mathbf{BlockAugment}_p(\, A_f(s,z)\,),\ \mathbf{L}(2+ \begin{bmatrix} zI_p \\ 0 \end{bmatrix} \times 1),\ \mathbf{R}(2+1\times(-\mathbf{A}(s)), \qquad (5.55)$$

where bold signs denote block counterparts of the scalar notation defined previously, the operator Augment is defined in (3.91) and

$$\mathbf{A}(s) := \begin{bmatrix} A_1(s) & 0 & 0 \\ 0 & A_2(s) & 0 \\ & & \ddots & \\ 0 & 0 & A_p(s) \end{bmatrix}. \qquad (5.56)$$

This yields the matrix

$$\begin{bmatrix} I_p & -\mathbf{A}(s) \\ zI_p & -\mathbf{B}_1(s) \\ 0 & -\mathbf{B}_2(s) \end{bmatrix} \qquad (5.57)$$

and then applying the procedure used in the previous cases gives the required system matrix as

$$\begin{bmatrix} sI_t - A & -B \\ \hline & z & 0 \\ -C & z & \ddots & -D \\ & 0 & z \\ & 0 & \cdots & 0 \end{bmatrix}_{q,p}. \qquad (5.58)$$

Again, it is necessary to avoid using all non-augmenting elementary operations (row- and column-) within the block

$$\mathbf{Z} \triangleq \begin{bmatrix} zI_p \\ 0 \end{bmatrix}$$

(now the structure of $\mathbf{Z}$ does not prevent illegal operations), and also row ones within

$$\begin{bmatrix} -\mathbf{B}_1(s) \\ -\mathbf{B}_2(s) \end{bmatrix}.$$

Example 5.10 Consider the same transfer function matrix as in Example 5.8. Start with the initial representation (5.53), apply the block step (5.57), and next apply the remaining steps of the procedure in Example 5.8 to finally obtain the system matrix

$$\begin{bmatrix} sI_4 - A & -B \\ \hline -C & \begin{bmatrix} z & 0 \\ 0 & z \end{bmatrix} & -D \end{bmatrix} = \begin{bmatrix} s & 0 & 1 & 0 & 0 & 0 \\ 0 & s & 0 & 1 & 0 & 0 \\ 0 & 0 & s+2 & 0 & 0 & 1 \\ 0 & 1 & 0 & s & 1 & 0 \\ \hline -2 & 1 & 1 & 1 & z & 0 \\ 0 & -1 & 1 & 1 & 0 & z \end{bmatrix},$$

where the A, B, C, D are the same as in Example 5.8.

At first sight, the representation of (5.53) appears to be the most appropriate basis for a further analysis. As noted previously, however, a major drawback is that the use of only one additional variable demands that great care must be exercised to avoid some elementary operations. For this reason the representation

$$A_f(s, z_1, \cdots, z_p)$$
$$= \begin{bmatrix} z_1 A_1(s) - B_{11}(s) & -B_{12}(s) & & -B_{1p}(s) \\ -B_{21}(s) & z_2 A_2(s) - B_{22}(s) & & -B_{2p}(s) \\ \vdots & \vdots & \cdots & \vdots \\ -B_{p1}(s) & -B_{p2}(s) & z_p A_p(s) - B_{pp}(s) \\ \vdots & \vdots & \cdots & \vdots \\ -B_{q1}(s) & -B_{q2}(s) & -B_{qp}(s) \end{bmatrix}. \tag{5.59}$$

has more desirable features. It is characterized by only a 'small' number of additional variables and it is only necessary to avoid using elementary operations in the last $q - p + 1$ rows, (the others are impossible because of the construction of (5.59)). The algorithm can be applied as before with the block step of (5.55) replaced by

$$\textbf{BlockAugment}_p(\,A_f(s, z_1, z_2, \dots, z_p)\,), \text{ L}(2+(-Z)\times 1), \text{ R}(2+1\times(A(s)), \tag{5.60}$$

where

$$\mathbf{Z} \triangleq \begin{bmatrix} z_1 & & & \\ & z_2 & & \\ & & \ddots & \\ & & & z_p \end{bmatrix}.$$

(5.61)

The resulting system matrix is now

$$\begin{bmatrix} sI_t - A & & -B & \\ \hline & \begin{bmatrix} z_1 & & & 0 \\ & z_2 & & \\ & & \ddots & \\ 0 & & z_p & \\ 0 & \cdots & & 0 \end{bmatrix}_{q,p} & -D \end{bmatrix}.$$

(5.62)

Case 2: $q < p$

Under this condition, there are a few ways of dealing with polynomial matrices using a limited number of additional variables. The first is to transpose the given transfer function matrix and then to work with a polynomial matrix of the form (5.53) or (5.59). The final system matrix resulting from application of the procedure (5.53)-(5.58) must then be transposed since

$$\left[D + C(sI_t - A)^{-1} B \right]^T = D^T + B^T \left(sI_t - A^T \right)^{-1} C^T.$$

(5.63)

The second approach is to combine the procedure of (5.33)-(5.36) with the one just described above. To describe this approach further, first write

$$F(s) = \left[f_{ij}(s) \right]_{q,p} = \left[F_1(s) \quad F_2(s) \quad \cdots \quad F_{k+1}(s) \right],$$

(5.64)

where $p = kq + r$ and $F_i(s) \in \mathrm{Mat}_{q,q}(R(s)), i = 1,2,\ldots,k$; $F_{k+1}(s) \in \mathrm{Mat}_{q,r}(R(s))$, $\Re(s)$ being the ring of rational functions in the complex variable s. In addition, the following notation is required

$$\mathbf{B}(s) = \left[b_{ij}(s) \right]_{q,p} = \left[\mathbf{B}_1(s) \quad \mathbf{B}_2(s) \quad \cdots \quad \mathbf{B}_{k+1}(s) \right],$$

(5.65)

where $\mathbf{B}_i(s) \in \mathrm{Mat}_{q,q}(R[s]), i = 1,2,\ldots,k$, $\mathbf{B}_{k+1}(s) \in \mathrm{Mat}_{q,r}(R[s])$ and $R[s]$ denotes, as usual, the ring of polynomials in s and

$$\mathbf{A}_j(s) \triangleq \begin{bmatrix} A_{1j}(s) & 0 & & 0 \\ 0 & A_{2j}(s) & & 0 \\ & & \ddots & \\ 0 & 0 & & A_{qj}(s) \end{bmatrix},$$ (5.66)

$j = 1,2,\ldots,k$, and

$$A_{k+1}(s) \triangleq \begin{bmatrix} A_{1,k+1}(s) & 0 & \cdots & 0 \\ 0 & A_{2,k+1}(s) & & \\ \vdots & & \ddots & \\ 0 & & & A_{r,k+1}(s) \\ & & \cdots & \\ & & 0 & \end{bmatrix}_{q,r},$$ (5.67)

where $A_{ij}(s), i = 1,2,\ldots,q; j = 1,2,\ldots,k+1$ denotes the least common denominator of all the entries in a given column of $F(s)$.

At this stage, one of two routes can be followed. The first is to introduce the polynomial matrix

$$\begin{aligned} A_f(s, z_1,\ldots,z_{k+1}) \\ = \begin{bmatrix} z_1 \mathbf{A}_1(s) - \mathbf{B}_1(s) & z_2 \mathbf{A}_2(s) - \mathbf{B}_2(s) & \cdots & z_{k+1} \mathbf{A}_{k+1}(s) - \mathbf{B}_{k+1}(s) \end{bmatrix} \end{aligned}$$ (5.68)

and the algorithm procedure can be applied in the similar manner to that described earlier, *i.e.*

BlockAugment$_n$($A_f(s, z_1,\cdots,z_{k+1})$), L($2+z_1 I_n \times 1$), R($2+1 \times (-\mathbf{A}_1(s))$) $\rightarrow A_{f1}$,

BlockAugment$_n$ (A_{f1}), L($3+z_2 I_n \times 1$), R($4+1 \times (-\mathbf{A}_2(s))$) $\rightarrow A_{f2}$,

$\cdots$ (5.69)

BlockAugment$_n$ ($A_{f,k-1}$), L($(k+1)+z_k I_n \times 1$), R($2k+1 \times \mathbf{A}_k(s)$) $\rightarrow A_{fk}$,

BlockAugment$_r$ (A_{fk}), L($(k+2)+ z_{k+1} E_{n,r} \times 1$), R($(2k+2)+1 \times \mathbf{A}_{k+1}(s)$) $\rightarrow A_{f,k+1}$,

where

$$E_{n,r} = \begin{bmatrix} I_r \\ 0 \end{bmatrix}_{n \times r}.$$ (5.70)

This yields

$$\begin{bmatrix} I_r & 0 & \cdots & 0 & 0 & 0 & 0 & \cdots & 0 & -A_{k+1}(s) \\ 0 & I_q & \cdots & 0 & 0 & 0 & 0 & \cdots & -A_k(s) & 0 \\ \vdots & \vdots & \ddots & \vdots & \vdots & \vdots & \vdots & \ddots & \vdots & \vdots \\ 0 & 0 & \cdots & I_q & 0 & 0 & -A_2(s) & \cdots & 0 & 0 \\ 0 & 0 & \cdots & 0 & I_q & -A_1(s) & 0 & \cdots & 0 & 0 \\ z_{k+1}E_{q,r} & z_k I_q & \cdots & z_2 I_q & z_1 I_q & -B_1(s) & -B_2(s) & \cdots & -B_k(s) & -B_{k+1}(s) \end{bmatrix}$$ (5.71)

and operating on this matrix as before, leads to the required system matrix

$$\left[\begin{array}{c|c} sI_t - A & -B \\ \hline -C & \begin{bmatrix} z_1 I_q & \cdots & z_k I_q & z_{k+1}E_{q,r} \end{bmatrix} - D \end{array}\right].$$ (5.72)

Note here that all elementary operations within the blocks $\begin{bmatrix} z_1 I_q & \cdots & z_k I_q & z_{k+1}E_{q,r} \end{bmatrix}$ are not allowed at each step of the algorithm, which prevents the use of row operations within $\begin{bmatrix} B_1(s) & \cdots & B_k(s) & B_{k+1}(s) \end{bmatrix}$. The following example illustrates this approach.

Example 5.11 Consider the transfer function matrix

$$\begin{bmatrix} \dfrac{s+1}{s+2} & \dfrac{1}{s+1} & \dfrac{1}{s} \\ \dfrac{s}{s-1} & \dfrac{s}{s-2} & \dfrac{1}{s+2} \end{bmatrix},$$

which can be written as

$$\begin{bmatrix} \dfrac{(s+1)(s-1)}{(s-1)(s+2)} & \dfrac{s-2}{(s-2)(s+1)} & \dfrac{s+2}{s(s+2)} \\ \dfrac{s(s+2)}{(s-1)(s+2)} & \dfrac{s(s+1)}{(s-2)(s+1)} & \dfrac{s}{s(s+2)} \end{bmatrix}.$$

Next, introduce the polynomial matrix (5.68), which in this case is given by

$$\begin{bmatrix} z_1(s-1)(s+2)-(s+1)(s-1) & -(s-2) & z_2 s(s+2)-s-2 \\ -s(s+2) & z_1(s-2)(s+1)-s(s+1) & -s \end{bmatrix}$$

and the matrix (5.72) is given by

$$A_{f1}(s,z_1,z_2) = \left[\begin{array}{ccc|ccc} 1 & 0 & 0 & 0 & 0 & -s(s+2) \\ 0 & 1 & 0 & -(s-1)(s+2) & 0 & 0 \\ 0 & 0 & 1 & 0 & -(s-2)(s+1) & 0 \\ \hline z_2 & z_1 & 0 & -s^2+1 & -s+2 & -s-2 \\ 0 & 0 & z_1 & -s^2-2s & -s(s+1) & -s \end{array}\right].$$

Now, apply

Augment(A_{f1}), L($3+1\times(s-1)$), L($5+1\times(s-2)$), L($6+1\times s$), R($5+1\times(s+2)$) $\rightarrow A_{f2}$,

Augment(A_{f2}), L($5+1\times(s-2)$, L($6+1\times1$), L($7+1\times s$), R($7+1\times(s+1)$) $\rightarrow A_{f3}$,

Augment(A_{f3}), L($4+1\times s$), L($7+1\times1$), L($8+1\times1$), R($9+1\times(s+2)$) $\rightarrow A_{f4}$,

L($7+5\times(-1)$), L($8+5\times(-1)$), L($8+6\times(-1)$).

Finally, applying appropriate row and column permutations to the result of these operations yields the system matrix

$$
\begin{bmatrix}
s+2 & 0 & 0 & 1 & 0 & 0 & 0 & 0 & 0 \\
0 & s+1 & 0 & 0 & 0 & 1 & 0 & 0 & 0 \\
0 & 0 & s+2 & 0 & 1 & 0 & 0 & 0 & 0 \\
0 & 0 & 0 & s & 0 & 0 & 0 & 0 & 1 \\
0 & 0 & 0 & 0 & s-1 & 0 & 1 & 0 & 0 \\
0 & 0 & 0 & 0 & 0 & s-2 & 0 & 1 & 0 \\
0 & 3 & -3 & 1 & -1 & 1 & z_1-1 & 0 & z_2 \\
2 & 0 & 0 & 1 & 1 & 2 & -1 & z_1-1 & 0
\end{bmatrix}.
$$

Clearly, in this example scalar (not-block) operations have been applied. As in the relevant previous cases, it is possible to allow use of some elementary operations at each step by using the initial representation

$$A_f(s, z_{11}, \ldots, z_{k+1,q})$$

$$= [\mathbf{Z}_1 \mathbf{A}_1(s) - \mathbf{B}_1(s) \quad \mathbf{Z}_2 \mathbf{A}_2(s) - \mathbf{B}_2(s) \quad \cdots \quad \mathbf{Z}_{k+1} \mathbf{A}_{k+1}(s) - \mathbf{B}_{k+1}(s)],$$

(5.73)

where

$$
\mathbf{Z}_i \triangleq \begin{bmatrix}
z_{i1} & & & \\
& z_{i2} & & \\
& & \ddots & \\
& & & z_{iq}
\end{bmatrix}
$$

(5.74)

for $i=1,2,\ldots,k$ and

$$
\mathbf{Z}_{k+1} \triangleq \begin{bmatrix}
z_{k+1,1} & & & \\
& \ddots & & \\
& & z_{k+1,r} & \\
& & & 0
\end{bmatrix},
$$

(5.75)

This form can be analysed as described previously, where the first stage operations are

$$\textbf{BlockAugment}_q(\,A_f(s,\mathbf{Z}_1,\cdots,\mathbf{Z}_{k+1})\,),\ \mathbf{L}(2+\mathbf{Z}_1 I_q\times 1),\ \mathbf{R}(2+1\times(-\mathbf{A}_1(s)))\ \rightarrow\ A_{f1},$$

$$\textbf{BlockAugment}_q(A_{f1}),\ \mathbf{L}(3+\mathbf{Z}_2\,I_q\times 1),\ \mathbf{R}(4+1\times(-\mathbf{A}_2(s)))\ \rightarrow\ A_{f2},$$

$$\cdots \tag{5.76}$$

$$\textbf{BlockAugment}_q\,(A_{f,k-1}),\ \mathbf{L}((k+1)+\mathbf{Z}_k\,I_q\times 1),\ \mathbf{R}(2k+1\times A_k(s))\ \rightarrow\ A_{fk},$$

$$\textbf{BlockAugment}_r(A_{fk}),\ \mathbf{L}((k+2)+\mathbf{Z}_{k+1}\,E_{q,r}\times 1),\ \mathbf{R}((2k+2)+1\times A_{k+1}(s))\ \rightarrow\ A_{f,k+1}.$$

This yields

$$\begin{bmatrix} I_r & 0 & \cdots & 0 & 0 & 0 & 0 & \cdots & 0 & -\mathbf{A}_{k+1}(s) \\ 0 & I_q & \cdots & 0 & 0 & 0 & 0 & \cdots & -\mathbf{A}_k(s) & 0 \\ \vdots & \vdots & \ddots & \vdots & \vdots & \vdots & \vdots & \iddots & \vdots & \vdots \\ 0 & 0 & \cdots & I_q & 0 & 0 & -\mathbf{A}_2(s) & \cdots & 0 & 0 \\ 0 & 0 & \cdots & 0 & I_q & -\mathbf{A}_1(s) & 0 & \cdots & 0 & 0 \\ \mathbf{Z}_{k+1} & \mathbf{Z}_k & \cdots & \mathbf{Z}_2 & \mathbf{Z}_1 & -\mathbf{B}_1(s) & -\mathbf{B}_2(s) & \cdots & -\mathbf{B}_k(s) & -\mathbf{B}_{k+1}(s) \end{bmatrix}. \tag{5.77}$$

and it is a straightforward exercise to obtain the required system matrix as

$$\left[\begin{array}{c|c} sI_t - A & -B \\ \hline -C & [\mathbf{Z}_1\ \cdots\ \mathbf{Z}_k\ \mathbf{Z}_{k+1}] - D \end{array}\right]. \tag{5.78}$$

In this approach the 'ban' on using elementary operations within the block $[\mathbf{Z}_1\ \cdots\ \mathbf{Z}_k\ \mathbf{Z}_{k+1}]$ is not required because such operations are not now permitted! This prevents using row operations within the block $[-\mathbf{B}_1(s)\ -\mathbf{B}_2(s)\ \cdots\ -\mathbf{B}_{k+1}(s)]$ and hence such an initial form does not require any additional assumptions as to which operations are allowed or banned.

It is also possible to apply here the dual approach outlined in Section 5.3 and generalised in (5.50)-(5.52). This is very similar to the previous relevant cases and hence the details are omitted here.

Example 5.12 Consider the same transfer function matrix as in Example 5.11, written in the form

$$\begin{bmatrix} \dfrac{(s+1)^2 s}{(s+1)(s+2)s} & \dfrac{(s+2)s}{(s+1)(s+2)s} & \dfrac{(s+1)(s+2)}{(s+1)(s+2)s} \\[2ex] \dfrac{s(s-2)(s+2)}{(s-1)(s-2)(s+2)} & \dfrac{s(s-1)(s+2)}{(s-1)(s-2)(s+2)} & \dfrac{(s-1)(s-2)}{(s-1)(s-2)(s+2)} \end{bmatrix}$$

and introduce the associated polynomial matrix

$$A_f = \begin{bmatrix} z_1(s+1)(s+2)s - (s+1)^2 s & -(s+2)s & -(s+2)(s+1) \\ z_2(s-1)(s+2)(s-2) - s(s+2)(s-2) & -s(s+2)(s-1) & -(s-1)(s-2) \end{bmatrix}.$$

Next apply the operations

```
> B1:=Augment(A_f);

> B1:=addcol(B1,1,2,z1);

> B1:=addrow(B1,1,2,-(s+1)*s*(s+2));

> B2:=Augment(B1);

> B2:=addcol(B2,1,3,z2);

> B2:=addrow(B2,1,4,-(s-1)*(s+2)*(s-2));

> B3:=Augment(B2);

> B3:=addrow(B3,1,4,o+1);

> B3:=addcol(B3,1,3,s*(s+2));

> B3:=addcol(B3,1,4,(s+1)*s);

> B3:=map(simplify,addcol(B3,1,5,s+1));

> B3:=addcol(B3,1,6,(s+2));

> B4:=Augment(B3);

> B4:=addrow(B4,1,6,s+2);

> B4:=addcol(B4,1,3,(s-2)*(s-1));

> B4:=addcol(B4,1,5,s*(s-2));

> B4:=addcol(B4,1,6,s*(s-1));

> B4:=map(simplify,addcol(B4,1,7,s-5));

> B5:=Augment(B4);

> B5:=addrow(B5,1,2,s-1);

> B5:=map(simplify,addcol(B5,1,4,-s+2));

> B5:=map(simplify,addcol(B5,1,6,-s+1));

> B5:=map(simplify,addcol(B5,1,7,-s));

> B5:=addcol(B5,1,8,-1);

> B6:=Augment(B5);

> B6:=addrow(B6,1,4,s+2);

> B6:=addcol(B6,1,6,-s);
```

```
> B6:=addcol(B6,1,7,-s+1);

> B6:=addcol(B6,1,8,-1);

> B6:=addcol(B6,1,9,-1);

> B6:=addcol(B6,5,7,-1);

> B6:=addcol(B6,5,8,-1);

> B6:=addcol(B6,6,7,-1);

> B6:=mulrow(B6,1,-1);

> B6:=mulrow(B6,2,-1);.
```

Finally, apply the following row and column permutations:

```
> pr:=[4,3,8,7,2,1,6,5]:       #permutation for rows

> pc:=[1,2,3,4,5,6,7,8,9]: #permutation for columns
```

to obtain the system matrix

$$
\begin{bmatrix}
s+2 & 0 & 0 & 1 & 0 & 0 & 2 & -1 & 0 \\
0 & s-1 & 1 & 0 & 0 & 0 & -1 & 0 & -4 \\
0 & 0 & s+2 & 0 & 0 & 0 & 0 & 0 & -12 \\
0 & 0 & 0 & s+1 & 0 & 0 & 0 & 1 & 0 \\
0 & -1 & 0 & 0 & s-2 & 0 & 1 & 2 & 1 \\
-1 & 0 & 0 & 0 & 0 & s & -1 & 1 & 1 \\
0 & 0 & 0 & 0 & 0 & 1 & z_1-1 & 0 & 0 \\
0 & 0 & 0 & 0 & 1 & 0 & z_2-1 & -1 & 0
\end{bmatrix}
$$

and this matrix has the same dimension as its counterpart in the previous example.

5.6 An Algebraic Characterization

The EOA for MIMO systems can be algebraically characterised in a similar manner to the SISO case (Gałkowski (1996)), as presented in Section 3.2 for bivariate transfer functions, where it is shown that the two different steps of the EOA produce zero equivalent polynomial matrices. The following is a similar result for the extensions of this algorithm developed here.

Theorem 5.1 *The matrices $A_{f_i}(s,\mathbf{z})$ and $A_{f_j}(s,\mathbf{z})$, where $\mathbf{z}$ denotes a collection of more than one additional variables that result from two different steps of the Elementary Operations Algorithm are zero equivalent.*

Proof It is easy to see from the algorithm that there exist 2-D unimodular matrices $P(s,\mathbf{z})$ and $Q(s,\mathbf{z})$ such that

$$\begin{bmatrix} I_\partial & 0 \\ 0 & A_{f_i}(s,z) \end{bmatrix} P(s,z) = Q(s,z) A_{f_j}(s,z), \tag{5.79}$$

where

$$A_{f_i}(s,z) \in \mathrm{Mat}_{n_1,m_1}(R[s,z]), \quad A_{f_j}(s,z) \in \mathrm{Mat}_{n_2,m_2}(R[s,z]),$$

$$P(s,z), \in \mathrm{Mat}_{m_2,m_2}(R[s,z]), \quad Q(s,z), \in \mathrm{Mat}_{n_2,n_2}(R[s,z]), \quad m_2 \geq m_1,$$

i.e. there exists a positive integer ∂ such that $m_2 = m_1 + \partial$. ∎

Let

$$P(s,z) = \begin{bmatrix} P_1(s,z) \\ T_2(s,z) \end{bmatrix}, Q(s,z) = \begin{bmatrix} Q_1(s,z) \\ T_1(s,z) \end{bmatrix}, \tag{5.80}$$

where

$$P_1(s,z) \in \mathrm{Mat}_{\partial,m_2}(R[s,z]), T_2(s,z) \in \mathrm{Mat}_{m_1,m_2}(R[s,z]),$$

$$Q_1(s,z) \in \mathrm{Mat}_{\partial,n_2}(R[s,z]), T_2(s,z) \in \mathrm{Mat}_{n_1,n_2}(R[s,z]),$$

Then the matrix equation (5.79) can be written as

$$P_1(s,z) = Q_1(s,z) A_{f_j}(s,z), \tag{5.81}$$

$$A_{f_i}(s,z) T_2(s,z) = T_1(s,z) A_{f_j}(s,z). \tag{5.83}$$

It is obvious that (5.82) establishes the equivalence relation between $A_{f_i}(s,z)$ and $A_{f_j}(s,z)$ and it remains to check zero coprimeness of the respective matrices.

a. **Left zero coprimeness.** Due to the unimodularity of $Q(s,z)$, a simple consequence of the Quillen-Suslin theorem, Gałkowski (1994), shows that $T_1(s,z)$ is zero coprime. Hence, the matrix $\begin{bmatrix} A_{f_i}(s,z) T_1(s,z) \end{bmatrix}$ is also zero coprime.

b. **Right zero coprimeness.** Due to the unimodularity of $P(s,z)$, the matrix $\begin{bmatrix} P(s,z) \\ A_{f_j}(s,z) \end{bmatrix}$ is zero coprime. Hence, there exist the polynomial matrices $U(s,z)$ and $V(s,z)$ such that the following Bezout identity holds:

$$U(s,z)P(s,z) + V(s,z) A_{f_j}(s,z) = I. \tag{5.83}$$

Also, note that it is possible to partition the matrix $U(s,z)$ as

$$U(s,\mathbf{z}) = \begin{bmatrix} U_1(s,\mathbf{z}) & U_2(s,\mathbf{z}) \end{bmatrix}$$ (5.84)

and such that

$$U(s,\mathbf{z})P(s,\mathbf{z}) = U_1(s,\mathbf{z})P_1(s,\mathbf{z}) + U_2(s,\mathbf{z})T_2(s,\mathbf{z}).$$ (5.85)

Substituting (5.84) into (5.82) and noting (5.81) yields

$$U_2(s,\mathbf{z})T_2(s,\mathbf{z}) + \begin{bmatrix} V(s,\mathbf{z}) + U_1(s,\mathbf{z})Q_1(s,\mathbf{z}) \end{bmatrix} A_{f_j}(s,\mathbf{z}) = I.$$ (5.86)

This is just the Bezout identity for the matrices $T_2(s,\mathbf{z})$ and $A_{f_j}(s,\mathbf{z})$ and hence they are zero right coprime. The key point here is that these results are also valid for the multi-dimensional case. Also, note that the algebraic motivation of the EOA can also be established in an equivalent simpler way, as detailed in the next chapter for 2D MIMO systems.

6. Multiple-input, Multiple-output (MIMO) Systems – the 2D Case

In this chapter the main objective is to extend the EOA case to the 2D MIMO case. Note, however, that the direct application of this method will almost certainly lead to a singular system as the final outcome. Here, this problem is solved by using variable transformations, Gałkowski (1992), (1999), (2000a).

The starting point is again the proper, rational matrix, which is now bivariate,

$$F(s_1, s_2) = \left[f_{ij}(s_1, s_2) \right] \; {}_{i=1,2,\ldots,q; \, j=1,2,\ldots,p}, \tag{6.1}$$

where

$$f_{ij}(s_1, s_2) = \frac{b_{ij}(s_1, s_2)}{a_{ij}(s_1, s_2)} \tag{6.2}$$

and $a_{ij}(s_1, s_2)$, $b_{ij}(s_1, s_2)$ are polynomials in s_1, s_2.

In the case of a 2D rational matrix, all the approaches detailed in the previous chapter for the 1D case are valid. In what follows, we first summarise the most useful of these and give a number of examples. Then, we proceed to consider the use of the variable inversion and generalised bilinear transform as techniques for avoiding unnecessary singularity.

6.1 The General Approach

Here we revisit two basic approaches based on:

a. using the least common multiple $A_j(s_1, s_2)$ of the denominators of all the elements in the jth column of the transfer matrix, and

b. using the least common multiple $A_i(s_1, s_2)$ of the denominators of all the elements in the ith row of the transfer matrix.

Also, the method based on using the least common multiple $A(s_1, s_2)$ of all the denominators of all the elements in the whole transfer matrix is briefly discussed.

6.1.1 Column Least Common Multiple Based Approach

The first step of this method is calculating the least common multiple $A_j(s_1,s_2)$ of all the denominators of all the elements in the jth column of the transfer matrix, *i.e.*

$$f_{ij}(s_1,s_2) = \frac{b_{ij}(s_1,s_2)}{a_{ij}(s_1,s_2)} = \frac{B_{ij}(s_1,s_2)}{A_j(s_1,s_2)}, \ i=1,2,\ldots,q, j=1,2,\ldots,p. \tag{6.3}$$

which is clearly equivalent to the polynomial matrix description

$$F(s_1,s_2) = N_R(s_1,s_2)D_R(s_1,s_2)^{-1} \tag{6.4}$$

with diagonal matrix $D_R(s_1,s_2)$.

Comment 6.1 Note here that it is also possible to construct a general coprime representation of (6.4) using the technique given in Guiver and Bose (1982). As shown in the previous chapter for the 1D case, however, application the EOA encounters severe difficulties. Note, however, that in such cases it should be possible to obtain a state-space description for the 2D case by 'pure' (*i.e.* non iterative) matrix methods (Bose (2000)).

In the previous chapter, results from Gałkowski (1997a) were used to develop various EOA procedures for 2D MIMO systems, which differ in the numbers of additional variables required. The simplest of these only requires one additional variable, as in the SISO 2D case, but certain forms of elementary operations are not permitted. Now, assume $n \geq m$ (the case of $m \geq n$ can, for example, be considered by first taking the transpose of the polynomial matrix involved) and consider the representation

$$A_f(s_1,s_2,z) = \begin{bmatrix} zI_p \\ 0 \end{bmatrix} D(s_1,s_2) - N(s_1,s_2)$$

$$= \begin{bmatrix} zA_1(s_1,s_2)-B_{11}(.) & -B_{12}(.) & -B_{1p}(.) \\ -B_{21}(.) & zA_2(.)-B_{22}(.) & -B_{2p}(.) \\ \vdots & \vdots & \cdots & \vdots \\ -B_{p1}(.) & -B_{p2}(.) & zA_p(.)-B_{pp}(.) \\ \vdots & \vdots & \cdots & \vdots \\ -B_{q1}(.) & -B_{q2}(.) & -B_{qp}(.) \end{bmatrix}. \tag{6.5}$$

Introduce the notation

$$\mathbf{B}(s_1,s_2) = \begin{bmatrix} B_{ij}(s_1,s_2) \end{bmatrix}_{q,p} = \begin{bmatrix} \mathbf{B}_1(s_1,s_2) \\ \mathbf{B}_2(s_1,s_2) \end{bmatrix} \begin{matrix} \}p \\ \}q-p \end{matrix} \tag{6.6}$$

and

$$\textbf{BlockAugment}_p(\Omega) := \begin{bmatrix} I_p & 0 \\ 0 & \Omega \end{bmatrix}.$$

(6.7)

Now, apply the following block versions of the operations defined in the previous section, *i.e.*

$$\textbf{BlockAugment}_p(\, A_f(s_1,s_2,z)\,),\ L(2{+}1{\times}(\begin{bmatrix} zE_p \\ 0 \end{bmatrix})),\ R(2{+}1{\times}(-A(s_1,s_2))),$$

where the bold symbols denote block counterparts of the scalar notation,

$$\mathbf{A}(s_1,s_2) := \begin{bmatrix} 0 & 0 & \cdots & A_p(s_1,s_2) \\ \vdots & \vdots & \ddots & \vdots \\ 0 & A_2(s_1,s_2) & \cdots & 0 \\ A_1(s_1,s_2) & 0 & \cdots & 0 \end{bmatrix},$$

(6.8)

and

$$E_p = \begin{bmatrix} 0 & 0 & \cdots & 0 & 1 \\ 0 & 0 & \cdots & 1 & 0 \\ \vdots & \vdots & \ddots & \vdots & \vdots \\ 0 & 1 & \cdots & 0 & 0 \\ 1 & 0 & \cdots & 0 & 0 \end{bmatrix}$$

(6.9)

This yields the matrix

$$\begin{bmatrix} I_p & -\mathbf{A}(s_1,s_2) \\ zE_p & -\mathbf{B}_1(s_1,s_2) \\ 0 & -\mathbf{B}_2(s_1,s_2) \end{bmatrix}$$

(6.10)

and then applying the procedure as for the SISO case gives the required system matrix as

$$\begin{bmatrix} s_1 I_{t_1} \oplus s_2 I_{t_2} - A & -B \\ -C & \begin{bmatrix} z & & & 0 \\ & z & & \\ & & \ddots & \\ 0 & & & z \\ 0 & \cdots & & 0 \end{bmatrix}_{q,p} & -D \end{bmatrix}.$$

(6.11)

Note that the column operations on $\begin{bmatrix} -B_1(s_1,s_2) \\ -B_2(s_1,s_2) \end{bmatrix}$ are admissible. It is easy to see that, as for the 1D case, the use of forbidden operations can be prevented by using additional variables in $\begin{bmatrix} zE_p \\ 0 \end{bmatrix}$. This is equivalent to using as the initial representation

$$A_f(s_1,s_2,z_{11},z_{12},\ldots,z_{1p}\ldots z_{qp})$$

$$= \begin{bmatrix} \cdots & \vdots \\ & z_{ij}A_j(s_1,s_2)-B_{ij}(s_1,s_2) & \cdots \\ & \vdots \end{bmatrix}_{i=1,2,\ldots,q;\,j=1,2\ldots p} \tag{6.12}$$

instead of (6.5) but then the number of additional variables is prohibitive. Clearly, all the other versions of the matrix $\begin{bmatrix} zE_p \\ 0 \end{bmatrix}$ detailed in the material devoted to 1D systems are valid here too. Next, the following example is given.

Example 6.1 Consider the transfer function matrix

$$F(s_1,s_2)= \begin{bmatrix} \dfrac{1}{s_1 s_2 +1} & \dfrac{1}{s_1 +1} \\[2mm] \dfrac{1}{s_2 -1} & \dfrac{1}{s_1 s_2 -1} \\[2mm] \dfrac{1}{(s_1 +1)(s_1 s_2 +1)} & \dfrac{1}{(s_2 -1)(s_1 s_2 -1)} \end{bmatrix},$$

which can be represented in the following form, *c.f.* (5.59),

$$A_f(s_1,s_2,z_1,z_2)= \begin{bmatrix} \begin{array}{c} z_1(s_1 s_2 +1)(s_2 -1)(s_1 +1) \\ -(s_2 -1)(s_1 +1) \end{array} & -(s_2 -1)(s_1 s_2 -1) \\ \hline \begin{array}{c} -(s_1 s_2 +1)(s_1 +1) \end{array} & \begin{array}{c} z_2(s_1 s_2 -1)(s_1 +1)(s_2 -1) \\ -(s_1 +1)(s_2 -1) \end{array} \\ \hline -(s_2 -1) & -(s_1 +1) \end{bmatrix}.$$

Next, apply appropriate elementary operations to transform A_f to the form of (5.57) or, for example, to (5.48), where

$$\begin{bmatrix} z_1 & z_2 & \cdots & z_p \\ 0 & 0 & \cdots & 0 \\ & \cdots & \cdots & \\ 0 & 0 & \cdots & 0 \end{bmatrix}$$

is replaced by

$$\begin{bmatrix} 0 & 0 & \cdots & z_1 \\ \vdots & \vdots & \ddots & \vdots \\ 0 & z_{p-1} & \cdots & 0 \\ z_p & 0 & \cdots & 0 \end{bmatrix}, i.e.$$

$$\mathbf{B} = \begin{bmatrix} 1 & 0 & 0 & -(s_1 s_2 - 1)(s_2 - 1)(s_1 + 1) \\ 0 & 1 & -(s_1 s_2 + 1)(s_2 - 1)(s_1 + 1) & 0 \\ 0 & z_1 & -(s_2 - 1)(s_1 + 1) & -(s_1 s_2 - 1)(s_2 - 1) \\ z_2 & 0 & -(s_1 s_2 + 1)(s_1 + 1) & -(s_2 - 1)(s_1 + 1) \\ 0 & 0 & -(s_2 - 1) & -(s_1 + 1) \end{bmatrix}.$$

Now, use the following operations, where again for notational simplification the substitutions $s_1 = s$ and $s_2 = p$ are used,

```
> B1:=Augment(B);
> B1:=addcol(B1,1,4,p-1);
> B1:=addrow(B1,1,3,(s*p+1)*(s+1));
> B1:=addrow(B1,1,4,(s+1));
> B1:=map(simplify,addrow(B1,1,5,(s^2+s)));
> B1:=addrow(B1,1,6,1);
> B2:=Augment(B1);
> B2:=addcol(B2,1,2,s+1);
> B2:=map(simplify,addrow(B2,1,4,-s*p-1));
> B2:=map(simplify,addrow(B2,1,5,-1));
> B2:=map(simplify,addrow(B2,1,6,-s));
> B3:=Augment(B2);
> B3:=addcol(B3,1,7,s+1);
> B3:=addrow(B3,1,4,(p-1)*(s*p-1));
> B3:=map(simplify,addrow(B3,1,6,(p^2-p)));
> B3:=map(simplify,addrow(B3,1,7,(p-1)));
```

```
> B3:=map(simplify,addrow(B3,1,8,1));

> B4:=Augment(B3);

> B4:=addcol(B4,1,2,p-1);

> B4:=map(simplify,addrow(B4,1,5,-s*p+1));

> B4:=map(simplify,addrow(B4,1,7,-p));

> B4:=map(simplify,addrow(B4,1,8,-1));

> B5:=Augment(B4);

> B5:=addcol(B5,1,4,s);

> B5:=addrow(B5,1,7,p);

> B5:=addrow(B5,1,9,1);

> B6:=Augment(B5);

> B6:=addcol(B6,1,3,p);

> B6:=addrow(B6,1,7,s);

> B6:=addrow(B6,1,9,1);

> B7:=Augment(B6);

> B7:=addcol(B7,1,11,p);

> B7:=addrow(B7,1,10,-p);

> B8:=Augment(B7);

> B8:=addcol(B8,1,2,p);

> B8:=addrow(B8,1,11,1);

> B9:=Augment(B8);

> B9:=addrow(B9,1,3,p);

> B9:=addcol(B9,1,13,-1);

> B10:=Augment(B9);

> B10:=addcol(B10,1,13,s+2);

> B10:=map(simplify,addrow(B10,1,14,s));

> B11:=Augment(B10);

> B11:=addcol(B11,1,2,s);

> B11:=addrow(B11,1,15,-1);

> B12:=Augment(B11);

> B12:=addrow(B12,1,3,s);
```

```
> B12:=addcol(B12,1,15,-1);
```

and apply the following row and column permutations:

```
> pr:=[1,4,2,3,8,10,11,13,5,6,7,9,12,14,15,16,17]:
#permutation for rows
```

```
> pc:=[2,5,3,1,11,16,12,7,6,4,9,10,15,8,14,13]:
#permutation for columns,
```

where 's' represents the variable 's_1' and 'p' represents the variable 's_2', to finally obtain the following singular system matrix of the Roesser form

$$\left[\begin{array}{cc|cccccc|cccccc|cc}
0 & 0 & 0 & 1 & 0 & 0 & 0 & 0 & 0 & 0 & 0 & 0 & -1 & 0 & 0 & 0 \\
0 & 0 & 0 & 0 & 0 & -1 & 0 & 0 & 0 & 1 & 0 & 0 & 0 & 0 & 0 & 0 \\
\hline
1 & 0 & s_1 & 0 & 0 & 0 & 0 & 0 & 0 & 0 & 0 & 0 & 0 & 0 & 0 & 0 \\
0 & 0 & 1 & s_1 & 0 & 0 & 0 & 0 & 0 & 0 & 0 & 0 & 2 & 0 & 0 & 0 \\
0 & 0 & 0 & 0 & s_1 & 0 & 0 & 0 & 0 & 0 & 0 & 0 & 0 & 1 & 0 & 0 \\
0 & 0 & 0 & 0 & 0 & s_1+1 & 0 & 0 & 0 & 0 & 0 & 1 & 0 & 0 & 0 & 0 \\
0 & 0 & 0 & 0 & 1 & 0 & s_1+1 & 0 & 0 & 0 & 0 & 0 & 0 & 0 & 0 & 0 \\
0 & 0 & 0 & 0 & 0 & 0 & 0 & s_1 & 0 & 0 & 1 & 0 & 0 & 0 & 0 & 1 \\
\hline
0 & 1 & 0 & 0 & 0 & 0 & 0 & 0 & s_2 & 0 & 0 & 0 & 0 & 0 & 0 & 0 \\
0 & 0 & 0 & 0 & 0 & 0 & 0 & 0 & 1 & s_2 & 0 & 0 & 0 & 0 & 0 & 0 \\
0 & 0 & 0 & 0 & 0 & 0 & 1 & 0 & 0 & 0 & s_2 & 0 & 0 & 0 & 0 & 0 \\
0 & 0 & 0 & 0 & 0 & 0 & 0 & 0 & 0 & 0 & 1 & s_2-1 & 0 & 0 & 0 & 0 \\
0 & 0 & 0 & 0 & 0 & 0 & 1 & 0 & 0 & 0 & 0 & 0 & s_2-1 & 0 & 0 & 0 \\
0 & 0 & 0 & 0 & -1 & 0 & 0 & 0 & 0 & 0 & 0 & 0 & 0 & s_2 & 1 & 0 \\
\hline
0 & 1 & 0 & 0 & -1 & -1 & 0 & 1 & 0 & 0 & 0 & 0 & 0 & 0 & z_1 & 0 \\
-1 & 0 & 0 & 0 & 0 & 0 & 0 & 0 & 0 & 0 & -1 & 0 & -1 & 1 & 0 & z_2 \\
0 & 0 & 0 & 0 & 0 & 0 & 1 & 0 & 0 & 0 & 0 & 1 & 0 & 0 & 0 & 0 \\
\end{array}\right]
$$

This is a particular form of the system matrix of the form of

$$\begin{bmatrix} S-A & -B \\ -C & Z-D \end{bmatrix}, \tag{6.13}$$

where

$$S = \begin{bmatrix} 0 & 0 & 0 \\ 0 & s_1 I_\alpha & 0 \\ 0 & 0 & s_2 I_\beta \end{bmatrix} \tag{6.14}$$

and

$$Z = \begin{bmatrix} z_1 & & & 0 \\ & z_2 & & \\ & & \ddots & \\ 0 & & & z_p \\ 0 & & \cdots & 0 \end{bmatrix}.$$

(6.15)

Clearly, the state-space quadruple $\{A,B,C,D\}$ defines a singular state-space realization of the Roesser form for a given transfer function matrix. Note that the matrix S can be written as

$$S = E\Lambda ,$$

(6.16)

which is consistent with the singular representation of the Roesser model (2.25), where

$$E = \begin{bmatrix} 0 & 0 \\ 0 & I_{\alpha+\beta} \end{bmatrix},$$

(6.17)

$$\Lambda = s_1 I_\chi \oplus s_2 I_\delta \oplus s_1 I_\alpha \oplus s_2 I_\beta .$$

(6.18)

Note here that the transfer function matrix does not depend on the terms $s_1 I_\chi \oplus s_2 I_\delta$. This is related to the fact that this realization is clearly redundant, *i.e.* non-minimal and the redundant terms can occur in both directions of information propagation, *i.e.* vertical and horizontal. Hence, equivalent state-space realizations of the same 2D system may differ even in the structure of the matrix Λ of (6.18). Thus, χ and δ may vary, but subject to the constraint

$$\chi + \delta = N - (\alpha + \beta) ,$$

(6.19)

where N is the dimension of the state-space realisation. It is also clear that the singular form of 2D systems given by (6.13)-(6.14) is the simplest for the analysis of such systems.

6.1.2 Row Least Common Multiple Based Approach

The generalisation of the dual method, where the least common denominator is taken from the transfer function matrix rows, is straightforward and hence the details are omitted here. Instead, we give the following example of the case where the number of columns is greater than the number rows.

Example 6.2 Consider the transfer function matrix

$$F(s_1, s_2) = \begin{bmatrix} \dfrac{s_1 + 1}{s_1 + 2} & \dfrac{1}{s_1 s_2 + 1} & \dfrac{1}{s_2} \\[2ex] \dfrac{s_2}{s_2 - 1} & \dfrac{s_1}{s_1 s_2 - 2} & \dfrac{1}{s_1 + 2} \end{bmatrix},$$

which can be represented in the form

$$A_f(s_1, s_2, z_1, z_2)$$

$$= \begin{bmatrix} \begin{matrix} z_1(s_1 s_2 + 1)s_2(s_1 + 2) \\ -(s_2 - 1)(s_1 + 2)s_2 \\ \hline z_2(s_1 s_2 - 2)(s_1 + 2)(s_2 - 1) \\ -(s_1 + 2)(s_1 s_2 - 2)s_2 \end{matrix} & \begin{matrix} -(s_1 + 2)s_2 \\ \\ -(s_1 + 2)(s_2 - 1)s_1 \end{matrix} & \begin{matrix} -(s_1 s_2 + 1)(s_1 + 2) \\ \\ -(s_1 s_2 - 2)(s_2 - 1) \end{matrix} \end{bmatrix}.$$

Now apply the following chain of operations:

```
> B1:=Augment(Fspz₁z₂);
> B1:=addcol(B1,1,2,z1);
> B1:=map(simplify,addrow(B1,1,2,-(s*p+1)*p*(s+2)));
> B2:=Augment(B1);
> B2:=addcol(B2,1,3,z2);
> B2:=map(simplify,addrow(B2,1,4,-(p-1)*(s+2)*(s*p-
  2)));
> B3:=Augment(B2);
> B3:=addrow(B3,1,4,s*p+1);
> B3:=addcol(B3,1,3,p*(s+2));
> B3:=addcol(B3,1,4,(s+1)*p);
> B3:=map(simplify,addcol(B3,1,5,1));
> B3:=addcol(B3,1,6,(s+2));
> B4:=Augment(B3);
> B4:=addrow(B4,1,6,s*p-2);
> B4:=addcol(B4,1,3,(s+2)*(p-1));
> B4:=map(simplify,addcol(B4,1,5,s*p+2*p));
> B4:=map(simplify,addcol(B4,1,6,s+2));
> B4:=map(simplify,addcol(B4,1,7,p-1));
> B5:=Augment(B4);
```

```
> B5:=addrow(B5,1,2,s+2);

> B5:=map(simplify,addcol(B5,1,4,-p+1));

> B5:=map(simplify,addcol(B5,1,6,-p));

> B5:=map(simplify,addcol(B5,1,7,-1));

> B5:=addcol(B5,4,6,-1);

> B5:=map(simplify,addcol(B5,5,6,-1));

> B6:=Augment(B5);

> B6:=addrow(B6,1,4,s+2);

> B6:=map(simplify,addcol(B6,1,6,-p));

> B6:=map(simplify,addcol(B6,1,9,-1));

> B7:=Augment(B6);

> B7:=addrow(B7,1,9,s);

> B7:=addcol(B7,1,4,-p);

> B7:=addcol(B7,1,9,-s);

> B8:=Augment(B7);

> B8:=addrow(B8,1,9,p);

> B8:=addcol(B8,1,6,-s);

> B8:=addcol(B8,1,10,2);

> B9:=Augment(B8);

> B9:=addrow(B9,1,7,p);

> B9:=addcol(B9,1,10,1);

> B10:=Augment(B9);

> B10:=addrow(B10,1,7,p);

> B10:=addcol(B10,1,13,-1);

> B11:=Augment(B10);

> B11:=addrow(B11,1,5,s);

> B11:=addcol(B11,1,13,1);

> B12:=Augment(B11);

> B12:=addcol(B12,1,3,p);

> B12:=addrow(B12,1,9,-1);

> B13:=Augment(B12);
```

```
> B13:=addcol(B13,1,5,p);
> B13:=addrow(B13,1,11,-1);
> B14:=Augment(B13);
> B14:=addcol(B14,1,4,s);
> B14:=addrow(B14,1,8,-1);
> B14:=mulrow(B14,7,-1);
> B14:=mulrow(B14,8,-1);
> B14:=mulrow(B14,9,-1);
> B14:=mulrow(B14,10,-1);
```

and apply the following row and column permutations:

```
> pr:=[4,5,6,1,7,11,12,16,2,3,8,9,10,15,14,13]:
#permutation for rows
> pc:=[1,2,3,4,12,10,9,8,6,5,11,14,13,7,15,16,17]:
#permutation for columns,
```

where again 's' represents the variable 's_1' and 'p' represents the variable 's_2', to obtain the following system matrix related to the singular Roesser model

$$
\left[\begin{array}{ccc|ccccc|cccccc|ccc}
0 & 0 & 0 & 1 & 0 & 0 & 0 & 0 & 0 & 0 & 0 & 0 & 0 & 0 & 0 & 1 & 0 \\
0 & 0 & 0 & 0 & 0 & 0 & 0 & 0 & 0 & 1 & 0 & 0 & 0 & 0 & 0 & 0 & -1 \\
0 & 0 & 0 & 0 & 0 & 0 & 0 & 0 & 1 & 0 & 0 & 0 & 0 & 0 & 1 & 0 & 0 \\
\hline
1 & 0 & 1 & s_1 & 0 & 0 & 0 & 0 & 0 & 0 & 0 & 0 & 0 & 0 & 0 & 0 & 0 \\
0 & 0 & 0 & 0 & s_1 & 0 & 0 & 0 & 0 & 0 & 0 & 0 & 0 & -1 & 0 & -2 & 0 \\
0 & 0 & -1 & 0 & 0 & s_1+2 & 0 & 0 & 0 & 0 & 1 & 0 & 0 & 0 & 0 & 0 & -1 \\
0 & -1 & 0 & 0 & 1 & 0 & s_1+2 & 0 & 0 & 0 & 0 & 0 & 0 & 0 & 0 & 1 & 0 \\
0 & 0 & 0 & 0 & 0 & 0 & 0 & s_1 & 0 & 0 & -2 & 0 & 0 & 0 & 0 & -4 & 0 \\
\hline
0 & 1 & 0 & 0 & 0 & 0 & 0 & 0 & s_2 & 0 & 0 & 0 & 0 & 0 & 0 & 0 & 0 \\
0 & 0 & 1 & 0 & 0 & 0 & 0 & 0 & 0 & s_2 & 0 & 0 & 0 & 0 & 0 & 0 & 0 \\
1 & 0 & 0 & 0 & 0 & 0 & 0 & -1 & 0 & 0 & s_2 & 0 & 0 & 0 & 0 & 0 & 0 \\
0 & 0 & 0 & 0 & 0 & 0 & -1 & 0 & 0 & 0 & 1 & s_2 & 0 & 0 & 0 & 0 & 1 \\
0 & 0 & 0 & 0 & 0 & -1 & 0 & 0 & 0 & 0 & 0 & 0 & s_2-1 & 0 & 1 & 1 & 0 \\
0 & 0 & 0 & 0 & 1 & 0 & 0 & 0 & 0 & 0 & 0 & 0 & 0 & s_2 & 0 & 1 & 0 \\
\hline
0 & 0 & 0 & 0 & 0 & 0 & 0 & 0 & 0 & 0 & 0 & 1 & 0 & 0 & z_1-1 & 0 & 0 \\
0 & 0 & 0 & 0 & 0 & 0 & 0 & 0 & 0 & 0 & 0 & 0 & 1 & 0 & z_2-1 & 0 & 0
\end{array}\right]
$$

This clearly is a special case of (6.13), with

$$Z = \begin{bmatrix} z_1 & 0 & \cdots & 0 \\ z_2 & 0 & \cdots & 0 \\ \vdots & \vdots & & \vdots \\ z_q & 0 & \cdots & 0 \end{bmatrix}.$$

This is again of the form of (6.16)-(6.18) and the equivalence analysis to the previous example is also valid here.

Both the examples presented in this chapter suggest that the EOA yields singular results. In the next section, we develop methods for avoiding this result. However, it is sometimes possible to achieve the standard solution directly.

6.1.3 A Standard (Nonsingular) Realization by Direct Application of the EOA

Next, we give some examples of the use of the EOA to directly produce some standard realizations of 2D transfer function matrices. We begin with a case where the entries in this matrix have the same denominator (which is clearly principal) and all numerators are also principal.

Example 6.3 Consider the transfer function matrix

$$F(s_1, s_2) = \begin{bmatrix} \dfrac{s_1 + 1}{s_1 s_2 + s_1 + s_2 - 1} & \dfrac{1}{s_1 s_2 + s_1 + s_2 - 1} \\ \dfrac{-1}{s_1 s_2 + s_1 + s_2 - 1} & \dfrac{s_2 + 1}{s_1 s_2 + s_1 + s_2 - 1} \end{bmatrix}.$$

Note that this transfer function matrix has the intrinsic feature that all its entries have the same denominator. The initial polynomial matrix of the form of (6.9) in this case is

$$A_f(s_1, s_2) = \begin{bmatrix} z(s_1 s_2 + s_1 + s_2 - 1) - (s_1 + 1) & -1 \\ 1 & z(s_1 s_2 + s_1 + s_2 - 1) - (s_2 + 1) \end{bmatrix}.$$

Now, apply the following operations:

```
> B1 := Augment ( A_f(s_1,s_2) ) :

> B1 := addrow (B1, 1, 2, -z) :

> B1 := addcol (B1, 1, 2, ( s_1 * s_2 + s_1 + s_2 -1) ) :

> B2 := Augment (B1) :

> B2 := addrow (B2, 1, 4, -z) :

> B2 := addcol (B2, 1, 4, ( s_1 * s_2 + s_1 + s_2 -1) ) :
```

```
> B3  := Augment(B2):

> B3  := addcol(B3,1,4,  S₁):

> B3  := addrow(B3,1,3,- S₂-1):

> B3  := addrow(B3,1,4,1):

> B3  := addcol(B3,1,4,1):

> B4  := Augment(B3):

> B4  := addcol(B4,1,6,  S₂):

> B4  := addrow(B4,1,3,- S₁-1):

> B4  := addrow(B4,1,6,1):

> B4  := addcol(B4,1,6,1):

> B4  := swaprow(B4,1,3):

> B4  := swapcol(B4,2,5):

> B4  := swapcol(B4,3,6):

> B4  := swapcol(B4,4,5):

> B4  := mulrow(B4,1,-1):

> B4  := mulrow(B4,4,-1):

> B4  := mulrow(B4,5,-1):

> B4  := mulrow(B4,6,-1);
```

to obtain the following companion matrix

$$
\Lambda - H = \left[
\begin{array}{cc|cc|cc}
s_1+1 & 0 & 2 & 0 & 0 & -1 \\
0 & s_1+1 & 0 & 1 & 0 & 0 \\
\hline
1 & 0 & s_2+1 & 0 & 0 & 0 \\
0 & 2 & 0 & s_2+1 & -1 & 0 \\
\hline
0 & 0 & 1 & -1 & z & 0 \\
-1 & -1 & 0 & 0 & 0 & z
\end{array}
\right],
$$

which is clearly standard (or nonsingular). The associated state space realisation of the Roesser type is

$$
A = \begin{bmatrix} -1 & 0 & -2 & 0 \\ 0 & -1 & 0 & -1 \\ -1 & 0 & -1 & 0 \\ 0 & -2 & 0 & -1 \end{bmatrix}, \quad B = \begin{bmatrix} 0 & 1 \\ 0 & 0 \\ 0 & 0 \\ 1 & 0 \end{bmatrix}, \quad C = \begin{bmatrix} 0 & 0 & -1 & 1 \\ 1 & 1 & 0 & 0 \end{bmatrix}, \quad D = 0.
$$

Note that in this example appropriate row and column permutations have been applied manually.

The following example demonstrates that it is not necessary for a transfer function matrix to have one common denominator to directly obtain a standard solution.

Example 6.4 Consider the transfer function matrix

$$F(s_1, s_2) = \begin{bmatrix} \dfrac{1}{s_1 s_2 + 1} & \dfrac{1}{s_1 + 1} \\ \dfrac{-1}{s_2 + 1} & \dfrac{1}{s_1 s_2 - 1} \end{bmatrix}.$$

The initial polynomial matrix of the form of (6.9) in this case is

$$A_f(s_1, s_2) = \begin{bmatrix} z(s_1 s_2 + 1)(s_2 + 1) - (s_2 + 1) & -(s_1 s_2 - 1) \\ (s_1 s_2 + 1) & z(s_1 s_2 - 1)(s_1 + 1) - (s_1 + 1) \end{bmatrix}.$$

Now, apply the following operations:

```
> B1:=Augment(AFsp):
> B1:=addrow(B1,1,3,z):
> B1:=addcol(B1,1,3,-(s*p-1)*(s+1));
> B2:=Augment(B1):
> B2:=addrow(B2,1,3,z):
> B2:=addcol(B2,1,3,-(s*p+1)*(p+1));
> B3:=Augment(B2):
> B3:=addcol(B3,1,5,(s*p-1)):
> B3:=addrow(B3,1,3,s+1):
> B3:=map(simplify,addrow(B3,1,4,1));
> B4:=Augment(B3):
> B4:=addcol(B4,1,5,s*p+1):
> B4:=addrow(B4,1,3,p+1):
> B4:=map(simplify,addrow(B4,1,6,-1));
> B5:=Augment(B4):
> B5:=addcol(B5,1,6,p):
> B5:=map(simplify,addrow(B5,1,2,-s)):
> B5:=map(simplify,addrow(B5,1,6,1));
```

```
> B6:=Augment(B5):

> B6:=addcol(B6,1,8,s):

> B6:=map(simplify,addrow(B6,1,4,-p)):

> B6:=map(simplify,addrow(B6,1,8,1));

> B6:=mulrow(B6,3,-1):

> B6:=mulrow(B6,4,-1).
```

Finally, apply the following row and column permutations:

```
> pr:=[1,3,6,2,4,5,7,8]:        #permutation for rows

> pc:=[8,2,4,7,1,3,5,6]: #permutation for columns,
```

where again 's' represents the variable 's_1' and 'p' represents the variable 's_2', to yield the following system matrix:

$$
\begin{bmatrix}
s_1 & 0 & 0 & 0 & 1 & 0 & 0 & 0 \\
0 & s_1 & 0 & -1 & 0 & -1 & 0 & 0 \\
0 & 0 & s_1+1 & 0 & 0 & 0 & 0 & 1 \\
0 & 1 & 0 & s_2 & 0 & 0 & 0 & 0 \\
1 & 0 & -1 & 0 & s_2 & 0 & 0 & 0 \\
0 & 0 & 0 & 0 & 0 & s_2+1 & 1 & 0 \\
0 & 1 & 1 & -1 & 0 & 0 & z & 0 \\
-1 & 0 & 0 & 0 & 1 & -1 & 0 & z
\end{bmatrix},
$$

which clearly results in a standard Roesser model. Unfortunately, however, this feature is not often present, and instead even transfer function matrix with the principal property leads to a singular state-space realization. Hence, we need to develop procedures, which allow us to avoid the undesirable singularity property of the final realization.

Now, note that each transfer function matrix can easily be transformed to one with a common denominator of the form

$$
f_{ij}(s_1,s_2) = \frac{b_{ij}(s_1,s_2)}{a_{ij}(s_1,s_2)} = \frac{B_{ij}(s_1,s_2)}{A(s_1,s_2)}, \quad i = 1,2,...,q, j = 1,2,...,p, \tag{6.20}
$$

where $A(s_1,s_2)$ is a common denominator of $F(s_1,s_2)$. The the EOA procedure can also be applied to this case.

Example 6.5 Consider the transfer function matrix of the previous example and represent it now as

$A_f(s_1, s_2)$

$$= \begin{bmatrix} z(s_1s_2+1)(s_1s_2-1)(s_1+1)(s_2+1) \\ -(s_1s_2-1)(s_1+1)(s_2+1) \\ \hline (s_1+1)(s_1s_2+1)(s_1s_2-1) \end{bmatrix} \begin{array}{c} -(s_1s_2+1)(s_1s_2-1)(s_2+1) \\ \hline z(s_1s_2-1)(s_1s_2+1)(s_1+1)(s_2+1) \\ -(s_1+1)(s_2+1)(s_1s_2+1) \end{array} \Bigg].$$

Next apply the following operations:

```
> B1:=Augment( A_f(s_1,s_2) );

> B1:=addrow(B1,1,3,z):

> B1:=addcol(B1,1,3,-(s^2*p^2-1)*(s+1)*(p+1));

> B2:=Augment(B1):

> B2:=addrow(B2,1,3,z):

> B2:=addcol(B2,1,3,-(s^2*p^2-1)*(s+1)*(p+1));

> B3:=Augment(B2):

> B3:=addcol(B3,1,5,(s*p-1)):

> B3:=addrow(B3,1,3,(s+1)*(p+1)*(s*p+1)):

> B3:=addrow(B3,1,4,(p+1)*(s*p+1)):

> B3:=map(simplify,B3);

> B4:=Augment(B3):

> B4:=addcol(B4,1,2,(p+1)*(s*p+1)):

> B4:=addrow(B4,1,4,-s-1):

> B4:=map(simplify,addrow(B4,1,5,-1));

> B5:=Augment(B4):

> B5:=addrow(B5,1,2,s*p+1):

> B5:=map(simplify,addcol(B5,1,3,-p-1));

> B6:=Augment(B5):

> B6:=addcol(B6,1,2,p):

> B6:=addrow(B6,1,3,-s);

> B7:=Augment(B6):

> B7:=addcol(B7,1,9,s):

> B7:=addrow(B7,1,5,-p):

> B7:=map(simplify,addrow(B7,1,9,p*(s+1)*(p+1)));
```

```
> B8:=Augment(B7):

> B8:=addcol(B8,1,2,p):

> B8:=addrow(B8,1,6,1):

> B8:=addrow(B8,1,10,-(s+1)*(p+1));

> B9:=Augment(B8):

> B9:=addrow(B9,1,11,s+1):

> B9:=addcol(B9,1,2,p+1):

> B9:=map(simplify,addcol(B9,1,10,-s^2*p^2+1)):

> B9:=map(simplify,addcol(B9,1,11,p+1)):

> B9:=addcol(B9,2,11,-1);

> B10:=Augment(B9):

> B10:=addcol(B10,1,11,s*p-1):

> B10:=map(simplify,addrow(B10,1,2,s*p+1)):

>B10:=map(simplify,addrow(B10,1,9,(s*p+1)*(s+1)
*(p+1))):

> B10:=map(simplify,addrow(B10,1,11,(s+1)*(p+1)));

> B11:=Augment(B10):

> B11:=addcol(B11,1,2,s*p+1):

> B11:=addrow(B11,1,3,-1):

> B11:=map(simplify,addrow(B11,1,10,-(s+1)*(p+1)));

> B12:=Augment(B11):

> B12:=addrow(B12,1,11,s+1):

> B12:=addcol(B12,1,2,p+1);

> B13:=Augment(B12):

> B13:=addcol(B13,1,4,s+1):

> B13:=map(simplify,addrow(B13,1,3,-p)):

> B13:=map(simplify,addrow(B13,1,14,-p-1));

> B14:=Augment(B13):

> B14:=addrow(B14,1,4,p):

> B14:=addcol(B14,1,2,1):

> B14:=addcol(B14,1,5,1);
```

```
> B15:=Augment(B14):

> B15:=addrow(B15,1,6,s):

> B15:=addcol(B15,1,16,-p);

> B15:=mulrow(B15,1,-1):

> B15:=mulrow(B15,11,-1):

> B15:=mulrow(B15,12,-1):

> B15:=mulrow(B15,15,-1):

> B15:=mulrow(B15,16,-1);

> B16:=Augment(B15):

> B16:=addcol(B16,1,4,p):

> B16:=addrow(B16,1,17,-1);

> B17:=Augment(B16):

> B17:=addcol(B17,1,9,s):

> B17:=addrow(B17,1,19,-1):

> B17:=mulrow(B17,18,-1).
```

Now, apply the following row and column permutations:

```
>pr:=[4,15,16,5,17,8,1,14,11,7,2,10,12,3,6,9,13,18,19]
:      #permutation for rows
```

```
>pc:=[1,2,6,8,14,3,9,12,19,4,5,11,13,18,7,10,15,16,17]
:      #permutation for columns,
```

where again 's' represents the variable 's_1' and 'p' represents the variable 's_2', to achieve the system matrix

$$
\begin{bmatrix}
0 & 0 & 0 & 1 & 0 & 0 & 0 & 0 & 0 & 1 & 1 & 0 & 0 & 0 & 0 & 0 & 0 & 0 & 0 \\
0 & 0 & 0 & 0 & 0 & 0 & 0 & 0 & -2 & 0 & 0 & 0 & 0 & 0 & 0 & 1 & 1 & 0 & 0 \\
0 & 0 & s_1+1 & 0 & 0 & 0 & 0 & 0 & 0 & 0 & 0 & 0 & 0 & 0 & 0 & 0 & 0 & 1 & 0 \\
0 & 0 & 0 & s_1+1 & 0 & 0 & 0 & 0 & 0 & 0 & 1 & 0 & 0 & 0 & 0 & 0 & 0 & 0 & 0 \\
0 & 0 & 0 & 0 & s_1+1 & 0 & 0 & 0 & 0 & 0 & 0 & 0 & 0 & 0 & 0 & 0 & 0 & 0 & -1 \\
0 & 0 & 0 & 1 & 0 & s_1 & 0 & 0 & 0 & 0 & 0 & 0 & 0 & -1 & 0 & 0 & 0 & 0 & 0 \\
1 & 0 & 0 & 0 & 0 & 0 & s_1 & 0 & 0 & 0 & 0 & 0 & 0 & 0 & 0 & 0 & 0 & 0 & 0 \\
0 & 0 & 0 & 0 & -1 & 0 & 0 & s_1 & 0 & 0 & 0 & 0 & -1 & 0 & 0 & 0 & 0 & 0 & 0 \\
0 & 0 & 0 & 0 & 0 & 0 & 0 & 0 & s_1 & 0 & 0 & 1 & 0 & 0 & 0 & 0 & 0 & 0 & 0 \\
0 & 0 & 0 & 1 & 0 & 0 & 0 & 0 & 0 & s_2 & 0 & 0 & 0 & 0 & 1 & 0 & 0 & 0 & 0 \\
0 & 1 & 0 & 0 & 0 & 0 & 0 & 0 & 0 & 0 & s_2 & 0 & 0 & 0 & 0 & 0 & 0 & 0 & 0 \\
0 & 0 & 0 & 0 & 0 & 0 & 0 & 0 & -1 & 0 & 0 & s_2 & 0 & 0 & 0 & 1 & 0 & 0 & 0 \\
0 & 0 & 0 & 0 & 0 & 0 & 0 & 1 & 0 & 0 & 0 & 0 & s_2 & 0 & 0 & 0 & 0 & 0 & 0 \\
0 & 0 & 0 & 0 & 0 & -1 & 0 & 0 & 0 & 0 & 0 & 0 & 0 & s_2 & 0 & 0 & 0 & 0 & 0 \\
0 & 0 & 1 & 0 & 0 & 0 & 0 & 0 & 0 & 0 & 0 & 0 & 0 & 0 & s_2+1 & 0 & 0 & 0 & 0 \\
0 & 0 & 0 & 0 & 0 & 0 & 1 & 0 & 0 & 0 & 0 & 0 & 0 & 0 & -1 & s_2+1 & 0 & 0 & 0 \\
0 & 0 & 0 & 0 & 0 & 0 & 0 & 0 & 0 & 0 & 0 & 0 & -1 & 0 & 0 & 0 & s_2+1 & 0 & 0 \\
0 & 1 & 0 & 0 & -1 & 0 & 0 & 0 & 0 & 0 & -1 & 0 & 0 & 0 & 0 & 0 & 0 & z & 0 \\
-1 & 0 & 0 & 0 & 0 & 0 & 1 & 0 & 0 & 0 & 0 & 0 & 0 & 0 & 0 & 0 & 0 & 0 & z
\end{bmatrix}
$$

This matrix is both singular and of 'very large' dimension (exceeding that of the previous example). It clearly demonstrates that we have to be very careful in applying various forms of the EOA since some of them may very well lead to solutions which, although correct, are highly unsatisfactory.

Note also that we can conjecture that direct application of the EOA to the 2D MIMO case will, in general, yield singular realizations. Only in very special circumstances, when there are well-defined links between the numerators and denominators of the entries occur, a standard realization can be obtained.

6.2 Algebraic Justification for the MIMO Case

In the previous chapter Gałkowski (1997a) the justification of the EOA procedure for the MIMO case has been provided in terms of zero prime equivalence for multidimensional polynomial matrix systems description. In what follows, we return to this problem and give a simpler characterisation, which in turn gives some insights into how to derive a range of equivalent state-space realizations.

In the SISO case the determinant was responsible for the application appropriate elementary operations due to its invariance under them. In the MIMO case, as noted previously, the determinant must not be used for this purpose. The following lemma is critical here.

Lemma 6.1 *Given the polynomial matrix*

$$X = \begin{bmatrix} X_{11} & X_{12} \\ X_{21} & X_{22} \end{bmatrix} \tag{6.21}$$

the generalised transfer matrix

$$T(X) := X_{22} - X_{21}(X_{11})^{-1}X_{12} \tag{6.22}$$

is invariant under an augmentation operation, and the following elementary operations:

    (i)   *column operations on*

$$X_1 := \begin{bmatrix} X_{11} \\ X_{21} \end{bmatrix}, \tag{6.23}$$

    i.e.

$$\tilde{X}_1 = X_1 V = \begin{bmatrix} X_{11} \\ X_{21} \end{bmatrix} V, \tag{6.24}$$

where V is unimodular,

    (ii)   *row operations on*

$$X^1 := \begin{bmatrix} X_{11} & X_{12} \end{bmatrix}, \tag{6.25}$$

i.e.

$$\tilde{X}^1 = W X^1 = W \begin{bmatrix} X_{11} & X_{12} \end{bmatrix}, \tag{6.26}$$

where W is unimodular,

    (iii)   *mixed column operations*

$$\tilde{X}_2 := \begin{bmatrix} \tilde{X}_{12} \\ \tilde{X}_{22} \end{bmatrix} = \begin{bmatrix} X_{12} \\ X_{22} \end{bmatrix} + \begin{bmatrix} X_{11} \\ X_{21} \end{bmatrix} \Omega, \tag{6.27}$$

    (iv)   *mixed row operations*

$$\tilde{X}^2 := \begin{bmatrix} \tilde{X}_{21} & \tilde{X}_{22} \end{bmatrix} = \begin{bmatrix} X_{21} & X_{22} \end{bmatrix} + Q \begin{bmatrix} X_{11} & X_{12} \end{bmatrix}, \tag{6.28}$$

where Q and Ω are some matrices of compatible dimensions.

Proof:

    (i)   Consider the effect of augmentation, *i.e.*

$$\tilde{X} := \begin{bmatrix} \tilde{X}_{11} & \tilde{X}_{12} \\ \tilde{X}_{21} & \tilde{X}_{22} \end{bmatrix} = \begin{bmatrix} \begin{bmatrix} 1 & 0 \\ 0 & X_{11} \\ [0 & X_{21}] \end{bmatrix} & \begin{bmatrix} 0 \\ X_{12} \\ X_{22} \end{bmatrix} \end{bmatrix}.$$

Then it is easy to see that

$$T(\tilde{X}) = \tilde{X}_{22} - \tilde{X}_{21}\left(\tilde{X}_{11}\right)^{-1}\tilde{X}_{12} = X_{22} - \begin{bmatrix} 0 & X_{21} \end{bmatrix}\begin{bmatrix} 1 & 0 \\ 0 & X_{11} \end{bmatrix}^{-1}\begin{bmatrix} 0 \\ X_{12} \end{bmatrix} = T(X).$$

(ii) and (iii)

$$T(\tilde{X}) = \tilde{X}_{22} - \tilde{X}_{21}\left(\tilde{X}_{11}\right)^{-1}\tilde{X}_{12} = X_{22} - X_{21}V\left(W\tilde{X}_{11}V\right)^{-1}WX_{12} = T(X).$$

(iv) $\quad T(\tilde{X}) = \tilde{X}_{22} - X_{21}(X_{11})^{-1}\tilde{X}_{12}$

$$= X_{22} + X_{21}\Omega - X_{21}(X_{11})^{-1}(X_{12} + X_{11}\Omega) = X_{22} - X_{21}(X_{11})^{-1}X_{12} = T(X).$$

(v) $\quad T(\tilde{X}) = \tilde{X}_{22} - \tilde{X}_{21}\left(X_{11}\right)^{-1}X_{12}$

$$= X_{22} + QX_{12} - (X_{21} + QX_{11})(X_{11})^{-1}X_{12} = X_{22} - X_{21}(X_{11})^{-1}X_{12} = T(X). \blacksquare$$

The following theorem shows that any multivariate polynomial matrix is invariant under all augmentations and elementary operations required to implement the EOA.

Theorem 6.1 *Augmentations and elementary operations applied to polynomial matrices during the EOA procedure leave a given transfer function matrix invariant (with respect to some constant).*

Proof Using appropriate column permutations the representation (6.5) can be rewritten in the form

$$\begin{bmatrix} I_p & -A(s_1,s_2) \\ zE_p & -B_1(s_1,s_2) \\ 0 & -B_2(s_1,s_2) \end{bmatrix} \Rightarrow G(s_1,s_2) = \begin{bmatrix} D_R(s_1,s_2) & I_p \\ \hline N_R(s_1,s_2) & \begin{bmatrix} zI_p \\ 0 \end{bmatrix} \end{bmatrix}.$$

Also, is easy to see that the generalised transfer matrix $T(X)$ of (6.22) for $G(s_1,s_2)$ is equal to $\begin{bmatrix} zI_p \\ 0 \end{bmatrix} - N_R(s_1,s_2)D_R^{-1}(s_1,s_2)$. Hence, all the operations of the type defined by Lemma 6.1 leave a transfer function matrix invariant. $\blacksquare$

This result clearly shows that it is possible to apply elementary operations to the columns of $\begin{bmatrix} -\mathbf{B}_1(s_1,s_2) \\ -\mathbf{B}_2(s_1,s_2) \end{bmatrix}$. A similar result occurs for the dual approach based on the transfer function matrix representation of the form $D_L^{-1}(s_1,s_2)N_L(s_1,s_2)$, where the row operations are allowed. This, in effect, considerably increases the possibility of deriving equivalent realizations.

6.3 Variable Transformations for the Elementary Operation Algorithm

Unfortunately, use of EOA procedures does not always produce standard (nonsingular) realizations - even for the class of principal transfer function matrices, of which includes Example 6.1 and 6.2 as special cases. To achieve this desirable result, additional relationships between the numerators and denominators of the matrix entries are required, which can be difficult to identify when applying EOA procedures. A necessary condition for an EOA procedure to produce standard (nonsingular) realizations is that all polynomials involved are principal. Note also that all denominators of the transfer function matrix entries must be principal for existence a Roesser state-space realization, which is standard.

The fact that it is possible to achieve essentially singular realizations starting from principal polynomial matrices is a considerable limitation of the method. It is, however, possible to improve the procedure in such a way as to obtain standard (nonsingular) solutions for a much wider classes of principal transfer function matrices than is possible with the basic version of the the EOA. The route, as detailed next, is the use of appropriate variable transformations which have a long history in systems theory - see for example, Fettweis (1992). In particular, inversion of variables and the generalised bilinear transform are employed.

6.3.1 Variable Inversion

The results developed in (Gałkowski (1992)) for n-variate transfer functions are easily generalised to the MIMO case, $i.e.$ to transfer function matrices. Consider the following polynomial matrix:

$$A_f(\Lambda,\wp) = \begin{bmatrix} E\Lambda - H_{11} & -H_{12} \\ -H_{21} & \wp - H_{22} \end{bmatrix},$$

(6.29)

where

$$\Lambda = s_1 I_{t_1} \oplus s_2 I_{t_2}$$

(6.30)

and $\wp$ is the matrix of additional variables (z_{ij}).

Theorem 6.2 *The singular system matrix of the form of (6.29) can be transformed, by inverting the variables s_1 and s_2, to the following standard (nonsingular) form:*

$$A'_f\left(\Lambda^{-1},\wp\right)=-\begin{bmatrix} \Lambda^{-1}-H_{11}^{-1}E & H_{11}^{-1}H_{12} \\ -H_{21}H_{11}^{-1}E & \wp-H_{22}+H_{21}H_{11}^{-1}H_{12} \end{bmatrix} \tag{6.31}$$

provided that the matrix H_{11} is nonsingular.

Proof : First, right multiply the matrix A_f given in (6.29) by the block matrix $\begin{bmatrix} \Lambda^{-1} & 0 \\ 0 & I \end{bmatrix}$. Next, apply the block elementary operation $L\left(2-H_{21}H_{11}^{-1}\times1\right)$ and finally, extract from the left side the block matrix $\begin{bmatrix} -H_{11} & 0 \\ 0 & I \end{bmatrix}$.   ∎

Obviously, this technique fails when the matrix H_{11} is singular.

A problem arises now in the sense that the variables s_1 and s_2 are interpreted as forward shift operators in the above analysis but most hardware and software implementations take s_i^{-1}, $i=1,2$ as delay units. Hence in this context use of the inverse variable transformation would produce the following acausal 'state' equation

$$\begin{bmatrix} x^h(i-1,j) \\ x^v(i,j-1) \end{bmatrix} = A\begin{bmatrix} x^h(i,j) \\ x^v(i,j) \end{bmatrix} + Bu(i,j). \tag{6.32}$$

This problem, however, can be avoided by applying the variable inversion transformation twice -first, to the initial transfer function matrix and then to the singular system matrix of the form (6.29) which results from application of the EOA procedure. On completing these, the original variables are recovered together with their mathematical and physical meanings.

Suppose now that (6.29) represents the singular system matrix associated with the Roesser model for the intermediately transformed transfer function. Then the final state space realisation of the original transform function is $\{A=H_{11}^{-1}E$, $B=-H_{11}^{-1}H_{12}$, $C=H_{21}H_{11}^{-1}E$, $D=H_{22}-H_{21}H_{11}^{-1}H_{12}\}$. Also, the dimension of this realisation can be reduced. In particular, assume that

$$E \triangleq \begin{bmatrix} 0I_\alpha & 0 \\ 0 & I_{t_1-\alpha+t_2} \end{bmatrix}. \tag{6.33}$$

Then, it is clear that

$$A = \begin{bmatrix} 0_{\alpha,\alpha} & A_{12} \\ 0 & \hat{A} \end{bmatrix}, \quad C = \begin{bmatrix} 0_{q,\alpha} & \hat{C} \end{bmatrix}. \tag{6.34}$$

Theorem 6.3 *The realization* $\{ A = H_{11}^{-1}E , \ B = -H_{11}^{-1}H_{12} , \ C = H_{21}H_{11}^{-1}E ,$ $D = H_{22} - H_{21}H_{11}^{-1}H_{12} \}$ *can be reduced to* $\{\hat{A}, \ \hat{B}, \ \hat{C}, \ D\}$ *of dimension* $t_1 + t_2 - \alpha$, *where*

$$B = \begin{bmatrix} *_{\alpha,p} \\ \hat{B} \end{bmatrix} \tag{6.35}$$

and * *denotes an arbitrary, appropriately dimensioned, matrix.*

Proof: Follows immediately from the fact that

$$F(s_1, s_2) = D + C\left[E\left(s_1 I_{t_1} \oplus s_1 I_{t_1}\right) - A\right]^{-1} B$$

$$= D + \begin{bmatrix} 0_{\alpha,\alpha} & | & \hat{C} \end{bmatrix} \begin{bmatrix} s_1^{-1}I_\alpha & | & * \\ \hline 0 & | & s_1 I_{t_1-\alpha} \oplus s_2 I_{t_2} - \hat{A} \end{bmatrix}^{-1} \begin{bmatrix} *_{\alpha,p} \\ \hat{B} \end{bmatrix}.$$

$$= D + \hat{C}\left[s_1 I_{t_1-\alpha} \oplus s_2 I_{t_2} - \hat{A}\right]^{-1} \hat{B}$$

The only problem remaining is under what circumstances is the matrix H_{11} nonsingular.

Theorem 6.4 *The sub-matrix* H_{11} *resulting from direct application of the EOA procedure to a 2D rational transfer function matrix is nonsingular if and only if the least common multiples of denominators for each matrix column has a nonzero free element, i.e. corresponding to the term* $s_1^{\,0}s_2^{\,0}$.

Proof. The proof is an immediate consequence of the operations, which implement the EOA procedure. ∎

This analysis leads immediately to the next, very important result.

Theorem 6.5 *The condition of Theorem 6.4 holds for the transformed transfer function matrix (with inverted variables) if and only if the original transfer function matrix is proper and principal, i.e. all its entries are principal. Moreover, all the entries of the associated polynomial matrix must be principal in the extended sense as detailed in Chapter 2.*

Proof The proof is an immediate consequence of applying the variable inversion.∎

Example 6.6 Consider the transfer function matrix

$$F(s_1,s_2) = \left[\begin{array}{c|c} \dfrac{s_1 s_2}{s_1 s_2 + 1} & \dfrac{s_1}{s_1 + 1} \\ \hline -\dfrac{s_2}{s_2 - 1} & -\dfrac{s_1 s_2}{s_1 s_2 - 1} \\ \hline \dfrac{s_1^2 s_2}{(s_1 s_2 + 1)(s_1 + 1)} & \dfrac{s_1 s_2^2}{(s_1 s_2 - 1)(s_2 - 1)} \end{array}\right].$$

Direct application of the EOA in this case yields a singular solution. Hence, invert its variables ($s_i := s_i^{-1}, i = 1,2$) to obtain the transfer function matrix, which turns out to be the matrix considered in Example 6.1. It is easy to check that the conditions of Theorem 6.4 hold and hence the resulting matrix H_{11} is nonsingular. Next, apply variable inversion to the resulting singular system matrix of this example. Finally, calculating the system matrix of (6.31) and applying the reduction procedure of Theorem 6.3 yields the following standard (nonsingular) state-space realization

$$\hat{A} = \begin{bmatrix} 0 & -1 & 0 & 0 & 2 & 0 & 0 & 0 & 0 & 0 & -2 & 2 \\ 0 & 0 & 0 & 0 & -1 & 0 & 0 & 0 & 0 & 0 & 1 & -1 \\ 0 & 0 & 0 & 0 & 0 & 0 & 0 & 0 & 0 & 0 & 0 & 1 \\ 0 & 0 & 0 & -1 & 0 & 0 & 0 & 0 & 0 & -1 & 0 & 0 \\ 0 & 0 & 0 & 0 & -1 & 0 & 0 & 0 & 0 & 0 & 0 & -1 \\ 0 & 0 & 0 & 0 & 0 & 0 & 0 & 0 & -1 & 0 & 0 & 0 \\ 0 & 0 & 0 & 0 & 0 & 0 & 0 & -1 & 0 & 0 & 0 & 0 \\ 0 & 0 & 0 & -1 & 0 & 1 & 0 & 0 & 0 & -1 & 0 & 0 \\ 0 & 0 & 0 & 0 & 0 & -1 & 0 & 0 & 0 & 0 & 0 & 0 \\ 0 & 0 & 0 & 0 & 0 & -1 & 0 & 0 & 0 & 1 & 0 & 0 \\ 0 & 0 & 0 & 0 & -1 & 0 & 0 & 0 & 0 & 0 & 1 & -1 \\ 0 & 0 & -1 & 0 & 0 & 0 & 0 & 0 & 0 & 0 & 0 & 0 \end{bmatrix}, \hat{B} = \begin{bmatrix} 2 & 0 \\ -1 & 0 \\ 1 & 0 \\ 0 & 1 \\ -1 & 0 \\ 0 & 0 \\ 0 & 0 \\ 0 & 1 \\ 0 & -1 \\ 0 & -1 \\ -1 & 0 \\ 0 & 0 \end{bmatrix},$$

$$\hat{C} = \begin{bmatrix} 0 & 0 & 0 & -1 & 0 & 1 & 1 & 0 & 1 & -1 & 0 & 1 \\ -1 & 0 & 1 & 0 & -1 & -1 & 0 & 0 & 0 & 0 & 1 & -1 \\ 0 & 0 & 0 & 0 & 1 & 1 & 0 & 0 & 0 & -1 & 0 & 1 \end{bmatrix}, D = \begin{bmatrix} 1 & 1 \\ -1 & -1 \\ 1 & 1 \end{bmatrix},$$

$$\Lambda = s_1 I_6 \oplus s_2 I_6.$$

Also, note that Theorem 6.3 is valid for partial variable inversion, i.e. in the case when we invert only one variable. Then,

$$\Lambda = s_1 I_{t_1} \text{ or } s_2 I_{t_2} \tag{6.36}$$

and $\wp$ is the matrix containing the additional variables and the second unchanged variable, e.g.

$$
\begin{bmatrix} s_2 I_{t_2} \text{ or } s_1 I_{t_1} & 0 \\ 0 & \begin{bmatrix} z E_m \\ 0 \end{bmatrix} \end{bmatrix}.
\tag{6.37}
$$

Additional comments are appropriate at this stage. Consider, for example, the transfer function matrix

$$
F(s_1, s_2) = \begin{bmatrix} \dfrac{s_1 + 1}{s_1 s_2 + 1} & \dfrac{1}{s_1 + 1} \\ \dfrac{1}{s_2 + 1} & \dfrac{s_2 - 1}{s_1 s_2 - 1} \end{bmatrix},
\tag{6.38}
$$

where the denominators of all entries are clearly principal, but inversion of variables leads to the matrix

$$
F'(s_1, s_2) = \begin{bmatrix} \dfrac{(s_1 + 1)s_2}{s_1 s_2 + 1} & \dfrac{s_1 s_2}{(s_1 + 1)s_2} \\ \dfrac{s_1 s_2}{s_1(s_2 + 1)} & -\dfrac{s_1(s_2 - 1)}{s_1 s_2 - 1} \end{bmatrix}
\tag{6.39}
$$

which does not satisfy the conditions of Theorem 6.4. However, if we reduce the number of variables in the numerator and the denominator in elements 12 and 21, the condition of Theorem 6.4 is satisfied and the method can be used. Hence, inversion of the variables leads to the nonsingular solution in all the cases when it exists.

We also provide an interesting example when, although the variable inversion results in a non-proper transfer function, the method works properly.

Example 6.7 Consider the transfer function matrix

$$
F(s_1, s_2) = \begin{bmatrix} \dfrac{s_1 + s_2 + 1}{s_1 s_2 + 1} & \dfrac{1}{s_1} \end{bmatrix}^T,
$$

for which application of the EOA procedure directly produces a singular result. Hence, invert both variables ($s_i := s_i^{-1}, i = 1,2$) to yield clearly a non-proper rational matrix

$$
F(s_1, s_2) = \begin{bmatrix} \dfrac{s_1 s_2 + s_1 + s_2}{s_1 s_2 + 1} & s_1 \end{bmatrix}^T.
$$

Now, the condition of Theorem 6.4 can be considered doubtful, but it is clear that the transfer function $s_1 = \dfrac{s_1}{1}$ has the denominator equal to 1, and hence its coefficient at $s_1^0 s_2^0$ is nonzero. The associated polynomial representation for $F(s_1, s_2)$ can be defined as

$$A_f(s_1, s_2, z_1, z_2) = \begin{bmatrix} z_1(s_1 s_2 + 1) - (s_1 s_2 + s_1 + s_2) \\ z_2 - s_1 \end{bmatrix}.$$

Apply hence the following operation chain

```
> B1:=Augment ( A_f (s_1,s_2,z_1,z_2) );
> B1:=addcol(B1,1,2,z1);
> B1:=map(simplify,addrow(B1,1,2,-(s*p+1)));
> B2:-Augment(B1);
> B2:=addcol(B2,1,3,z2);
> B2:=addrow(B2,1,4,-1);
> B3:=Augment(B2);
> B3:=addrow(B3,1,4,s);
> B3:=addcol(B3,1,3,p);
> B3:=map(simplify,addcol(B3,1,4,p+1));
> B3:=addrow(B3,1,4,1);
> B4:=Augment(B3);
> B4:=addcol(B4,1,4,p);
> B4:=addrow(B4,1,2,-1);
> B4:=addrow(B4,1,5,-1);
> B5:=Augment(B4);
> B5:=addrow(B5,1,7,s);
> B5:=addcol(B5,1,6,1);
> B6:=Augment(B5);
> B6:=addrow(B6,1,4,p);
> B6:=addcol(B6,1,7,-1).
```

Now, apply the following row and column permutations:

```
> pr:=[1,2,7,8,3,4,6,5]:      #permutation for rows
> pc:=[3,5,4,2,6,1,7]: #permutation for columns
```

to obtain the singular system matrix

$$\begin{bmatrix} 0 & 0 & 0 & 0 & 0 & 1 & -1 \\ 0 & 0 & 0 & 1 & 0 & 0 & 1 \\ -1 & 0 & s_1+1 & 0 & -1 & 0 & 1 \\ 0 & -1 & -0 & s_1 & 0 & 0 & 0 \\ 1 & 0 & 0 & 0 & s_2 & 0 & 0 \\ -1 & 0 & 1 & 0 & 0 & s_2 & 1 \\ 0 & 0 & 0 & 0 & 1 & 0 & z_1 \\ 0 & 1 & 0 & 0 & 0 & 0 & z_2 \end{bmatrix}.$$

Applying the variable inversion and reduction procedure of Theorem 6.3 now leads to the following standard (nonsingular) state-space realization of the original transfer function matrix:

$$A = \begin{bmatrix} 0 & 0 & -1 & 1 \\ 0 & 0 & 0 & 0 \\ 1 & 0 & 0 & -1 \\ 0 & 0 & 0 & 0 \end{bmatrix}, \quad B = \begin{bmatrix} -1 \\ -1 \\ 0 \\ 1 \end{bmatrix}, \quad C = \begin{bmatrix} -1 & 0 & 0 & 1 \\ 0 & -1 & 0 & 0 \end{bmatrix}, \quad D = \begin{bmatrix} 0 \\ 0 \end{bmatrix},$$

$$\Lambda = s_1 I_2 \oplus s_2 I_2, \quad E = I_4.$$

6.3.2 Generalised Bilinear Transform

An alternative to variable inversion, use of the generalised bilinear transform is considered in this section. This transform applied to the singular state-space realization can yield a standard (nonsingular) realization, Gałkowski (1992a), (2000a). However, use of the multivariate generalized bilinear transform

$$s_i \Rightarrow \frac{a_i s_i - b_i}{c_i s_i - a_i}$$ changes also the physical meaning of the system mathematical

model but in a more complex way than the inverse transform. If the variable transform is to be used, the route is to apply it twice - once to the initial transfer function matrix and then to the singular system matrix which results from applying the EOA procedure. On completing these, the original variables are recovered together with their mathematical and physical meanings.

Next we detail this for the bilinear transform.

The generalized multivariate bilinear transform is defined by the following pair of substitutions:

$$s_i = \frac{a_i z_i + b_i}{c_i z_i - a_i}, \tag{6.40}$$

and

$$z_i = \frac{a_i s_i + b_i}{c_i s_i - a_i}; \quad a_i, b_i, c_i \in R^+, \quad i = 1,2, \tag{6.41}$$

where s_i, z_i, $i = 1,2$ belong to the complex number field. Applying (6.40) to the transfer function matrix $F(s_1, s_2)$ yields

$$\tilde{F}(z_1, z_2) = F\left(\frac{a_1 z_1 + b_1}{c_1 z_1 - a_1}, \frac{a_2 z_2 + b_2}{c_2 z_2 - a_2}\right), \tag{6.42}$$

where, as a result, all the entries in the matrix F are reduced to polynomial form.

Theorem 6.6 *Application of the bilinear transform in the context of (6.42) can be expressed in state-space realization terms (Gałkowski (1992a)) as*

$$A' = -[E\Lambda_b + \Lambda_a A][E\Lambda_a - \Lambda_c A]^{-1}, \quad B' = (A'\Lambda_c - \Lambda_a)B,$$

$$C' = C[E\Lambda_a - \Lambda_c A]^{-1}, \quad D' = D + C'\Lambda_c B, \tag{6.43}$$

where

$$\Lambda_a := \bigoplus_{i=1}^{2} a_i I_{t_i}, \quad \Lambda_b := \bigoplus_{i=1}^{2} b_i I_{t_i}, \quad \Lambda_c := \bigoplus_{i=1}^{2} c_i I_{t_i}, \tag{6.44}$$

E is of the form (6.33) and the matrix $E\Lambda_a - \Lambda_c A$ is nonsingular.

Proof First, apply the bilinear transform in the form (6.40) to the singular system matrix

$$A_f^{1}(\Lambda, \wp) = \begin{bmatrix} E\Lambda - A & -B \\ -C & \wp - D \end{bmatrix}, \tag{6.45}$$

and left-multiply the resulting matrix by

$$\underbrace{(c_1 z_1 - a_1) \oplus \cdots \oplus (c_1 z_1 - a_1)}_{t_1} \oplus \underbrace{(c_2 z_2 - a_2) \oplus \cdots \oplus (c_2 z_2 - a_2)}_{t_2} \oplus I_n. \tag{6.46}$$

This yields a polynomial matrix of the form

$$A_f^{2}(\Lambda, \wp) = \begin{bmatrix} E(\Lambda_a \Lambda_z + \Lambda_b) - (\Lambda_c \Lambda_z - \Lambda_a)A & -(\Lambda_c \Lambda_z - \Lambda_a)B \\ -C & \wp - D \end{bmatrix}, \tag{6.47}$$

where

$$\Lambda_z := \bigoplus_{i=1}^{2} z_i I_{t_i}. \tag{6.48}$$

In what follows, note that the matrix E of the form of (6.33) commutes with diagonal matrices, which allows us to rewrite (6.47) in the form

$$A_f^{\,3}(\Lambda,\wp)=\begin{bmatrix} \Lambda_z(E\Lambda_a-\Lambda_c A)+(E\Lambda_b+\Lambda_a A) & -\Lambda_z\Lambda_c+\Lambda_a B \\ -C & \wp-D \end{bmatrix}. \tag{6.49}$$

Now, right multiply the resulting matrix by $\begin{bmatrix} (E\Lambda_a-\Lambda_c A)^{-1} & 0 \\ 0 & I_p \end{bmatrix}$ and apply the

block elementary operation $\mathbf{R}(\,A_f^{\,3}(\Lambda,\wp),1,2,\Lambda_c)$. This completes the proof. ∎

Alternatively, using column multiplications of the matrix (6.45) by the term

(6.46) and then left multiplication by $\begin{bmatrix} (E\Lambda_a-A\Lambda_c)^{-1} & 0 \\ 0 & I_q \end{bmatrix}$ leads after some

manipulations to the state-space realization of the form

$$A''=-[E\Lambda_a-A\Lambda_c]^{-1}[E\Lambda_b+A\Lambda_a],\quad B''=(E\Lambda_a-A\Lambda_c)B,$$

$$C''=C(\Lambda_a-\Lambda_c A''),\quad D''=D+C\Lambda_c B''. \tag{6.50}$$

Note here that this last approach is more general than the previous one since it does not impose any assumptions on the matrix E.

Now, the following algorithm for avoiding singularity can be stated.

a. Apply the bilinear transform of (6.41) to the entries in the given transfer function matrix.

b. Construct a state space realization for the transfer function matrix resulting from the previous step.

c. Use (6.43) or (6.50) to return to the original variables and hence recover the state space realization for the original transfer function matrix.

The following theorem now be stated as the counterpart to Theorem 6.3.

Theorem 6.7 *In the singular state-space realization* $\{A',B',C',D'\}$ *of the form (6.43) or (6.50) the first α columns of the matrices A and C and the first α rows of the matrices A and B can be removed without changing the final state space realization provided that (6.33) holds.*

Proof Consider the transfer function matrix resulting from application of the procedure given above

$$F(s_1,s_2)=D'+C'\big[\!\big(s_1 I_{t_1}\oplus s_2 I_{t_2}\big)-A'\big]^{-1}B'$$

and introduce

$$G=\big(s_1 I_{t_1}\oplus s_2 I_{t_2}\big)-A' \tag{6.51}$$

and

$$X = G^{-1}. \tag{6.52}$$

Then the transfer function matrix can be rewritten as

$$F(s_1, s_2) = D' + C'XB', \tag{6.53}$$

where $C'XB'$ can be rewritten in the block form

$$C'XB' = \begin{bmatrix} C_1' & C_2' \end{bmatrix} \begin{bmatrix} X_{11} & X_{12} \\ X_{21} & X_{22} \end{bmatrix} \begin{bmatrix} B_1' \\ B_2' \end{bmatrix} \tag{6.54}$$

and C_1, X_{11}, B'_1 are the $q \times \alpha$, $\alpha \times \alpha$, $\alpha \times p$ submatrices, respectively. It is straightforward to show that $C'XB'$ depends only on C_2, X_{22}, and B'_2, i.e.

$$C'XB' = C_2' X_{22} B_2' \tag{6.55}$$

independently of the matrix X if, and only if, $C_2' = 0$, $B_2' = 0$.

In what follows, we need to specify the conditions when the matrix X does not depend on the choice of the submatrix G_{11} of its inversion, i.e.

$$G = X^{-1} = \begin{bmatrix} G_{11} & G_{12} \\ G_{21} & G_{22} \end{bmatrix}. \tag{6.56}$$

Clearly, if G_{12}=0 then

$$X = G^{-1} = \begin{bmatrix} G_{11}^{-1} & 0 \\ -G_{22}^{-1}G_{21}G_{11}^{-1} & G_{22}^{-1} \end{bmatrix}. \tag{6.57}$$

and

$$C'XB' = C_2 G_{22}^{-1} B_2. \tag{6.58}$$

It is straightforward to show that G_{12}=0 if and only if the corresponding block of the transformed matrix $A' = \begin{bmatrix} A_{11}' & A_{12}' \\ A_{21}' & A_{22}' \end{bmatrix}$, i.e. A'_{12} is zero. Now, introduce

$$W = \begin{bmatrix} E\Lambda_a - \Lambda_c A \end{bmatrix}, \quad V = \begin{bmatrix} E\Lambda_b + \Lambda_a A \end{bmatrix} \tag{6.59}$$

and note that (6.43) is equivalent to

$$A'W = -V. \tag{6.60}$$

It is easy to conclude from (6.59) and (6.33) that both the matrices W and V can be written in the block form

$$W = \begin{bmatrix} 0 & W_{12} \\ W_{21} & W_{22} \end{bmatrix}, \quad V = \begin{bmatrix} 0 & V_{12} \\ V_{21} & V_{22} \end{bmatrix} \tag{6.61}$$

with compatible block dimensions, where

$$W_{12} = kV_{12} \tag{6.62}$$

and

$$k = \frac{c_1}{a_1} \text{ or } \frac{c_2}{a_2}. \tag{6.63}$$

In turn, this guarantees that A'_{12} is zero and $C'_2 = 0$, $B'_2 = 0$, which completes the proof.   ∎

This fact can also be established for (6.47).

To illustrate the above results, the following examples based on Maple V are now given.

Example 6.8 Consider the 2D transfer function matrix

$$F(z_1, z_2) = \begin{bmatrix} \dfrac{3z_1 z_2 - z_1 - z_2 - 3}{3z_1 z_2 + z_1 - z_2 + 3} & \dfrac{z_1 - 2}{3z_1 - 1} \\ -0.5\dfrac{z_2 - 1}{z_2} & \dfrac{(z_1 - 2)(z_2 - 1)}{z_1 z_2 + 3z_1 + 3z_2 - 1} \end{bmatrix}$$

and note that although this matrix is principal in the sense defined earlier in this chapter, direct application of the EOA procedure produces only singular realizations. Hence, introduce the following representation in the polynomial matrix form (6.5):

$$A_f(z_1, z_2) = \left[\begin{array}{c|c} \begin{array}{c} z(3z_1 z_2 + z_1 - z_2 + 3)z_2 \\ -(3z_1 z_2 - z_1 - z_2 - 3)z_2 \\ \hline 0.5(3z_1 z_2 - z_1 - z_2 - 3)(z_2 - 1) \end{array} & \begin{array}{c} -(z_1 z_2 + 3z_1 + 3z_2 - 1)(z_1 - 2) \\ \hline z(z_1 z_2 + 3z_1 + 3z_2 - 1)(3z_1 - 1) \\ -(3z_1 - 1)(z_1 - 2)(z_2 - 1) \end{array} \end{array} \right].$$

and apply to this matrix the following EOA procedure in the form of a string of elementary and augmentation operations:

```
> B1:=Augment ( A_f(z_1,z_2)) :

> B1:=addrow(B1,1,3,z) :

> B1:=addcol(B1,1,3,-(z1*z2+3*z1+3*z2-1)*(3*z1-1));

> B2:=Augment(B1) :

> B2:=addrow(B2,1,3,z) :

> B2:=addcol(B2,1,3,-(3*z1*z2+z1-z2+3)*z2);

> B3:=Augment(B2) :

> B3:=addcol(B3,1,5,(z1*z2+3*z1+3*z2-1)) :
```

```
> B3:=addrow(B3,1,3,3*z1-1):

> B3:=map(simplify,addrow(B3,1,4,z1-2));

> B4:=Augment(B3):

> B4:=addcol(B4,1,5,(3*z1*z2+z1-z2+3)):

> B4:=addrow(B4,1,3,z2):

> B4:=map(simplify,addrow(B4,1,6,-1/2*(z2-1))):

> B4:=addrow(B4,3,6,1/2):

> B4:=addrow(B4,4,5,-1/3);

> B5:=Augment(B4):

> B5:=addcol(B5,1,7,z2-1):

> B5:=map(simplify,addrow(B5,1,7,(3*z1-1)*(z1 2))),

> B6:=Augment(B5):

> B6:=addcol(B6,1,7,3*z1*z2-z1-z2-3):

> B6:=map(simplify,addrow(B6,1,3,-1)):

> B6:=map(simplify,addrow(B6,1,7,z2));

> B7:=Augment(B6):

> B7:=addcol(B7,1,9,z2-1):

> B7:=addrow(B7,1,3,-1):

> B7:=map(simplify,addrow(B7,1,5,-z1-3));

> B7:=addcol(B7,1,9,4);

> B8:=Augment(B7):

> B8:=addcol(B8,1,9,z1-1/3):

> B8:=map(simplify,addrow(B8,1,3,-3*z2)):

> B8:=addrow(B8,1,3,1):

> B8:=addrow(B8,1,5,-2);

> B9:=Augment(B8):

> B9:=addcol(B9,1,5,z1-2):

> B9:=map(simplify,addrow(B9,1,11,-(3*z1-1)));

> B10:=Augment(B9):

> B10:=addcol(B10,1,2,z1):

> B10:=addrow(B10,1,12,3);
```

```
> B10:=mulrow(B10,5,-1/3):
> B10:=mulrow(B10,8,-1):
> B10:=mulrow(B10,10,1/3);
> B11:=Augment(B10):
> B11:=addcol(B11,1,6,z2):
> B11:=addrow(B11,1,12,-1).
```

Now, apply appropriate row and column permutations to obtain the system matrix in the form

$$
\begin{bmatrix}
0 & 0 & 0 & 1 & 0 & -1 & 0 & 0 & -4 & 0 & 0 & 0 & 0 \\
0 & 0 & 0 & 0 & 20/3 & 0 & 0 & -1 & 0 & -2 & 1 & 0 & 0 \\
0 & 1 & z_1 & 0 & 0 & 0 & 0 & 0 & 0 & 0 & 0 & 0 & 0 \\
0 & 0 & 1 & z_1-2 & 0 & 0 & 0 & 0 & 0 & 0 & 0 & 0 & 0 \\
0 & 0 & 0 & 0 & z_1-1/3 & 0 & 0 & 0 & 0 & 1 & 0 & 0 & 0 \\
0 & 0 & 0 & 0 & 0 & z_1+3 & -1 & 0 & 10 & 0 & 0 & 0 & 0 \\
0 & 0 & 0 & 0 & 0 & 0 & z_1-1/3 & 0 & 0 & 0 & 0 & 0 & 1/3 \\
1 & 0 & 0 & 0 & 0 & 0 & 0 & z_2 & 0 & 0 & 0 & 0 & 0 \\
0 & 0 & 0 & 0 & 0 & 1 & 0 & 0 & z_2+3 & 0 & 0 & 0 & 0 \\
0 & 0 & 0 & 0 & 10/9 & 0 & 0 & -1/3 & 0 & z_2-1/3 & 0 & 0 & 0 \\
0 & 0 & 0 & 0 & 0 & 0 & 0 & 0 & 0 & 0 & z_2 & 1 & 0 \\
-1 & 0 & 0 & 0 & 0 & 0 & -5/3 & 0 & 0 & 0 & 0 & z & -1/3 \\
0 & 3 & 1 & 0 & 0 & 0 & 0 & 0 & 0 & 0 & 1/2 & 1/2 & z
\end{bmatrix},
$$

which is associated with a singular state space realisation of the form

$$
A_1 = \begin{bmatrix}
0 & 0 & 0 & -1 & 0 & 1 & 0 & 0 & 4 & 0 & 0 \\
0 & 0 & 0 & 0 & -20/3 & 0 & 0 & 1 & 0 & 2 & -1 \\
0 & -1 & 0 & 0 & 0 & 0 & 0 & 0 & 0 & 0 & 0 \\
0 & 0 & -1 & 2 & 0 & 0 & 0 & 0 & 0 & 0 & 0 \\
0 & 0 & 0 & 0 & 1/3 & 0 & 0 & 0 & 0 & -1 & 0 \\
0 & 0 & 0 & 0 & 0 & -3 & 1 & 0 & -10 & 0 & 0 \\
0 & 0 & 0 & 0 & 0 & 0 & 1/3 & 0 & 0 & 0 & 0 \\
-1 & 0 & 0 & 0 & 0 & 0 & 0 & 0 & 0 & 0 & 0 \\
0 & 0 & 0 & 0 & 0 & -1 & 0 & 0 & -3 & 0 & 0 \\
0 & 0 & 0 & 0 & -10/9 & 0 & 0 & 1/3 & 0 & 1/3 & 0 \\
0 & 0 & 0 & 0 & 0 & 0 & 0 & 0 & 0 & 0 & 0
\end{bmatrix},
$$

$$
B_1 = \begin{bmatrix} 0 & 0 \\ 0 & 0 \\ 0 & -1 \\ 0 & 0 \\ 0 & 0 \\ 0 & 0 \\ 0 & -1/3 \\ 0 & 0 \\ 0 & 0 \\ 0 & 0 \\ -1 & 0 \end{bmatrix}, C_1 = \begin{bmatrix} 1 & 0 & 0 & 0 & 0 & 5/3 & 0 & 0 & 0 & 0 \\ 0 & -3 & -1 & 0 & 0 & 0 & 0 & 0 & 0 & -1/2 \end{bmatrix},
$$

$$
D_1 = \begin{bmatrix} 0 & 1/3 \\ -1/2 & 0 \end{bmatrix}, \quad E\Lambda = 0I_2 \oplus z_1 I_5 \oplus z_2 I_4.
$$

Example 6.9 Consider application of the bilinear transform defined by

$$
s_1 = \frac{2z_1 + 1}{z_1 - 2}, \quad s_2 = \frac{z_2 + 1}{z_2 - 1},
$$

to the transfer function matrix from the previous example. This results in

$$
F(s_1, s_2) = \begin{bmatrix} \dfrac{s_1 + s_2}{s_1 s_2 + 1} & \dfrac{1}{s_1 + 1} \\ -\dfrac{1}{s_2 + 1} & \dfrac{1}{s_1 s_2 - 1} \end{bmatrix},
$$

and the initial polynomial matrix is

$$
A_f(s_1, s_2, w_1, w_2)
$$
$$
= \begin{bmatrix} w_1(s_1 s_2 + 1)(s_2 + 1) - (s_1 + s_2)(s_2 + 1) & -(s_1 s_2 - 1) \\ s_1 s_2 + 1 & w_2(s_1 s_2 - 1)(s_1 + 1) - (s_1 + 1) \end{bmatrix}.
$$

Now, apply the EOA procedure to this matrix:

```
> B1:=Augment(Af);
> B1:=addrow(B1,1,2,w1);
> B1:=map(simplify,addcol(B1,1,2,-(1+p)*(s*p+1)));
> B2:=Augment(B1):
> B2:=addrow(B2,1,4,w2):
```

```
> B2:=map(simplify,addcol(B2,1,4,-(s*p-1)*(s+1)));

> B3:=Augment(B2);

> B3:=addcol(B3,1,4,p+1);

> B3:=addrow(B3,1,3,s*p+1);

> B3:=map(simplify,addrow(B3,1,4,s+p));

> B3:=map(simplify,addrow(B3,1,5,-s));

> B4:=Augment(B3);

> B4:=addcol(B4,1,6,s+1);

> B4:=addrow(B4,1,3,s*p-1);

> B4:=map(simplify,addrow(B4,1,5,p));

> B4:=addrow(B4,1,6,1);

> B5:=Augment(B4);

> B5:=addcol(B5,1,2,p);

> B5:=map(simplify,addrow(B5,1,4,-s));

> B5:=addrow(B5,1,6,-1);

> B6:=Augment(B5);

> B6:=addcol(B6,1,4,s);

> B6:=map(simplify,addrow(B6,1,6,-p));

> B6:=map(simplify,addrow(B6,1,7,-1));

> B6:=addrow(B6,1,8,1);

> B6:=addcol(B6,4,8,-1);

> B6:=addrow(B6,3,1,1);

> B7:=Augment(B6);

> B7:=addcol(B7,1,5,p);

> B7:=addrow(B7,1,8,-1);

> B8:=Augment(B7);

> B8:=addcol(B8,1,9,s);

> B8:=map(simplify,addrow(B8,1,10,1));

> B9:=Augment(B8);

> B9:=addrow(B9,1,3,p);

> B9:=map(simplify,addcol(B9,1,7,-1));
```

```
> B10:=Augment(B9);
> B10:=addrow(B10,1,3,s);
> B10:=addcol(B10,1,11,-1);
> B10:=mulrow(B10,9,-1):
> B10:=mulrow(B10,10,-1);
```

(here again $s_1 := s$ and $s_2 := p$). Now, apply the following row and column permutations:

```
> pr:=[1,2,3,5,7,9,4,6,8,10,11,12]:          #permutation
for rows
> pc:=[3,4,1,8,12,6,2,7,11,5,10,9]:  #permutation   for
columns
```

to obtain the singular system matrix

$$
\begin{bmatrix}
0 & 0 & 1 & 0 & 0 & 0 & 0 & -1 & 0 & 0 & 0 & 0 \\
0 & 0 & 0 & -1 & 0 & 0 & 1 & 0 & 0 & 0 & 0 & 0 \\
1 & 0 & s_1 & 0 & 0 & 0 & 0 & 0 & 0 & 0 & 0 & 0 \\
0 & 0 & 0 & s_1 & 1 & 0 & 0 & 1 & 0 & 1 & 0 & 0 \\
0 & 0 & 0 & 0 & s_1+1 & 0 & 0 & 1 & 0 & 0 & 0 & 0 \\
0 & 0 & 0 & 0 & 0 & s_1 & 0 & 1 & 0 & 0 & 0 & -1 \\
0 & 1 & 0 & 0 & 0 & 0 & s_2 & 0 & 0 & 0 & 0 & 0 \\
0 & 0 & 0 & 0 & 0 & 1 & 0 & s_2 & 0 & 0 & 0 & 0 \\
0 & 0 & 0 & 0 & -1 & 0 & 0 & 0 & s_2+1 & 0 & 0 & 0 \\
0 & 0 & 0 & -1 & 1 & 0 & 0 & 0 & 0 & s_2 & -1 & 0 \\
0 & -1 & 0 & 0 & 1 & -1 & 0 & 0 & 0 & -1 & z_1 & 0 \\
1 & 0 & 0 & 0 & 0 & 0 & 0 & 1 & 1 & 1 & 0 & z_2
\end{bmatrix}.
$$

Here, to avoid a misunderstanding, the additional variables are termed w_1 and w_2.

Finally, applying the inverse bilinear transform as in (6.43) and the reduction procedure of Theorem 6.7 yields the following final standard (nonsingular) Roesser state-space realization:

$$A = \begin{bmatrix} 2 & 5/6 & -5/6 & 0 & 0 & 0 & -5/2 & -5/6 \\ 0 & 1/3 & 0 & -5/3 & 0 & 10/3 & 0 & 5/3 \\ 0 & 0 & 1/3 & -5/3 & 0 & 10/3 & 0 & 0 \\ 0 & 0 & 0 & -3 & 0 & 5 & 0 & 0 \\ 0 & -2/3 & 0 & -2/3 & 1 & 4/3 & 0 & 2/3 \\ 0 & 0 & 0 & 2 & 0 & -3 & 0 & 0 \\ 0 & 1/3 & -1/3 & 0 & 0 & 0 & 0 & -1/3 \\ 0 & -2/3 & 2/3 & 0 & 0 & 0 & 0 & -1/3 \end{bmatrix},$$

$$B = \begin{bmatrix} -5/6 & 0 \\ 5/3 & -5/3 \\ 0 & -5/3 \\ 0 & -5 \\ 2/3 & -2/3 \\ 0 & 2 \\ -1/3 & 0 \\ -4/3 & 0 \end{bmatrix}, \quad C = \begin{bmatrix} 0 & 0 & -2/3 & 1/3 & 1 & 1/3 & 0 & 1 \\ -1 & -1/2 & 1/2 & 1 & 0 & -2 & 1/2 & -1/2 \end{bmatrix},$$

$$D = \begin{bmatrix} 1 & 1/3 \\ -1/2 & 1 \end{bmatrix}, \quad E = I_8, \quad \Lambda = z_1 I_4 \oplus z_2 I_4.$$

Example 6.10 In this example, we compare the result of the last example to that obtained by variable inversion. The starting point is the matrix of Example 6.8, which after inverting both variables takes the form

$$F(z_1, z_2) = \begin{bmatrix} -\dfrac{3z_1 z_2 + z_1 + z_2 - 3}{3z_1 z_2 - z_1 + z_2 + 3} & \dfrac{2z_1 - 1}{z_1 - 3} \\ 0.5(z_2 - 1) & -\dfrac{(2z_1 - 1)(z_2 - 1)}{z_1 z_2 - 3z_1 - 3z_2 - 1} \end{bmatrix}.$$

Next, introduce the following representation in the polynomial matrix form (5.59)

$$A_f(z_1, z_2)$$
$$= \begin{bmatrix} 2w_1(3z_1 z_2 - z_1 + z_2 + 3) & -(z_1 z_2 - 3z_1 - 3z_2 - 1)(2z_1 - 1) \\ \quad + 2(3z_1 z_2 + z_1 + z_2 - 3) & \\ \hline -(3z_1 z_2 - z_1 + z_2 + 3)(z_2 - 1) & w_2(z_1 z_2 - 3z_1 - 3z_2 - 1)(z_1 - 3) \\ & \quad + (z_1 - 3)(2z_1 - 1)(z_2 - 1) \end{bmatrix}.$$

and apply to this polynomial matrix the following EOA procedure in the form of a sequence of elementary and augmentation operations (where again $z_1 := s$ and $z_2 := p$)

```
> B1:=Augment(AFz1z2);

> B1:=addrow(B1,1,2,w1);

>               B1:=map(simplify,addcol(B1,1,2,-2*(3*z1*z2-
z1+z2+3)));

> B2:=Augment(B1);

> B2:=addrow(B2,1,4,w2);

>       B2:=map(simplify,addcol(B2,1,4,-(z1*z2-3*z1-3*z2-
1)*(z1-3)));

> B3:=Augment(B2);

> B3:=addcol(B3,1,4,z2-1);

> B3:=map(simplify,addrow(B3,1,3,(z1*6+2)));

> B3:=map(simplify,addrow(B3,1,4,(-6*z1-2)));

> B3:=map(simplify,addrow(B3,1,5,(3*z1*z2-z1+z2+3)));

> B4:=Augment(B3);

> B4:=addcol(B4,1,6,z2-1);

> B4:=map(simplify,addrow(B4,1,3,(z1-3)^2));

> B4:=map(simplify,addrow(B4,1,5,(z1-3)*(2*z1-1)));

> B4:=map(simplify,addrow(B4,1,6,-(2*z1^2-7*z1+3)));

> B5:=Augment(B4);

> B5:=addcol(B5,1,2,z1-3);

> B5:=map(simplify,addrow(B5,1,4,(-z1+3)));

> B5:=map(simplify,addrow(B5,1,6,-(2*z1-1)));

> B5:=map(simplify,addrow(B5,1,7,(2*z1-1)));

> B6:=Augment(B5);

> B6:=addcol(B6,1,4,z1+1/3);

> B6:=map(simplify,addrow(B6,1,6,-6));

> B6:=map(simplify,addrow(B6,1,7,6));

> B6:=map(simplify,addrow(B6,1,8,-(3*z2-1)));

> B7:=Augment(B6);
```

```
> B7:=addcol(B7,1,9,z1);

> B7:=map(simplify,addrow(B7,1,6,-2*z1+2));

> B7:=map(simplify,addrow(B7,1,8,-4*z1-6));

> B7:=addcol(B7,3,1,-2);

> B7:=addrow(B7,6,8,-2);

> B7:=addrow(B7,6,9,2);

> B7:=addrow(B7,7,8,2);

> B8:=Augment(B7);

> B8:=addrow(B8,1,8,z1);

> B8:=map(simplify,addcol(B8,1,9,4));

> B9:=Augment(B8);

> B9:=addcol(B9,1,3,z1);

> B9:=map(simplify,addrow(B9,1,11,4));

> B10:=Augment(B9);

> B10:=addcol(B10,1,5,z2);

> B10:=addrow(B10,1,12,3);

> B10:=mulrow(B10,9,-1);

> B11:=Augment(B10);

> B11:=addrow(B11,1,5,z1);

> B11:=addcol(B11,1,13,-1).
```

Now, apply the following row and column permutations:

```
>                     pr:=[1,4,3,5,6,7,10,11,2,8,9,12,13]:
#permutation for rows

>   pc:=[2,3,5,1,9,8,7,4,6,13,12,11,10]:    #permutation
for columns
```

to yield the singular system matrix

$$
\begin{bmatrix}
0 & 0 & 0 & 1 & 0 & 0 & 0 & 0 & 0 & -1 & 0 & 0 & 0 \\
0 & 0 & 0 & 0 & 0 & 0 & 0 & 1 & 0 & 0 & 4 & 0 & 0 \\
0 & 1 & z_1 & 0 & 0 & 0 & 0 & 0 & 0 & 0 & 0 & 0 & 0 \\
0 & 0 & 1 & z_1 & 0 & 0 & 0 & 0 & 0 & 0 & 0 & 0 & 0 \\
0 & 0 & 0 & 0 & z_1+1/3 & 0 & 0 & 0 & 1 & 0 & 0 & 0 & 0 \\
0 & 0 & -2 & 0 & 0 & z_1-3 & 1 & 0 & 0 & 0 & 0 & 0 & 0 \\
0 & 0 & 4 & 0 & 0 & 0 & z_1-3 & 0 & 0 & 12 & 0 & 0 & -1 \\
0 & 0 & 0 & 0 & 0 & 0 & 0 & z_1-6 & 0 & -8 & 1 & 0 & 0 \\
1 & 0 & 0 & 0 & 0 & 0 & 0 & 0 & z_2 & 0 & 0 & 0 & 0 \\
0 & 0 & 0 & 0 & 0 & 1 & 0 & 0 & 0 & z_2-1 & 0 & 0 & 0 \\
0 & 0 & 0 & 0 & 1 & 0 & 0 & 0 & 0 & -1 & z_2-1 & 0 & 0 \\
0 & 0 & 0 & 0 & 0 & 0 & -5 & 0 & -6 & 20 & -20 & \omega_1+2 & -2 \\
3 & 4 & -6 & 0 & 10/3 & 0 & 5 & 0 & 1 & -24 & 0 & 0 & \omega_2+2
\end{bmatrix}.
$$

Finally, calculating the matrix of (6.31) and applying the reduction procedure of Theorem 6.3 yields the following standard (nonsingular) state-space realization:

$$
A =
\begin{bmatrix}
0 & -1 & 0 & 0 & 0 & 0 & 0 & 0 & 0 \\
0 & -2/3 & 0 & -1 & -1/3 & 0 & 0 & -3 & 0 \\
0 & 0 & 1 & 0 & 0 & 1/6 & 0 & 0 & -4/3 \\
0 & -2/3 & 0 & -1 & -1/3 & 0 & 0 & -4 & 0 \\
0 & -4 & 0 & -4 & -1 & 0 & 0 & -12 & 0 \\
0 & 0 & -4 & 0 & 0 & -2/3 & 0 & 0 & 4/3 \\
0 & 0 & -4/3 & 0 & 0 & -1/18 & 0 & 0 & 4/9 \\
0 & -2/3 & 0 & -1 & -1/3 & 0 & 0 & -3 & 0 \\
0 & 0 & 1 & 0 & 0 & 1/6 & 0 & 0 & -1/3
\end{bmatrix},
$$

$$
B =
\begin{bmatrix}
0 & 0 \\
0 & 1/3 \\
1/6 & 0 \\
0 & 1/3 \\
0 & 1 \\
-2/3 & 0 \\
-1/18 & 0 \\
0 & 1/3 \\
1/6 & 0
\end{bmatrix},
\quad
C =
\begin{bmatrix}
0 & -20/3 & 12 & 0 & 5/3 & 3 & 0 & 0 & -4 \\
4 & -2 & -2 & -4 & -3 & -1/2 & 3 & -12 & 4
\end{bmatrix},
$$

$$D = \begin{bmatrix} 1 & 1/3 \\ -1/2 & 1 \end{bmatrix}, \quad E = I_9, \quad \Lambda = z_1 I_6 \oplus z_2 I_3.$$

In this particular case the bilinear transform provides a slightly better result in the sense that its dimension is one less in comparison with the case when variable inversion was.

In summary, this chapter has developed a number of EOA based procedures have been developed for obtaining a range of state-space realizations of a given 2D transfer function matrix. These are very significant since, as discussed in Section 6.2.2, similarity (unlike the 1D case) does not provide a full class of state-space realizations. This, in turn, can lead to the best solution (*i.e.* structure optimisation). Note, however, that the EOA procedures can produce realizations of different dimensions – the aim, therefore, should be to use the ones which produce the least dimension.

7. The Elementary Operation Approach Extensions

In this chapter we consider some extensions of EOA based procedures. First, we consider the case of nD ($n>2$) systems and then proceed to consider repetitive processes and Laurent polynomials are respectively.

7.1 The nD Elementary Operation Algorithm

First, note that all EOA based methods and procedures considered in this work also hold for the nD ($n>2$) case. Hence, we do not give a comprehensive reworking here, but instead concentrate on the basic starting point - the problem of finding the companion matrix for a multivariate polynomial or polynomial matrix.

Note here that, in general, the step from 2D to nD ($n>2$) is not trivial. In particular, for the 1D case all coprimeness properties are the same, *i.e.* factor (F), minor (M), and zero (Z) coprimeness. For the 2D case, only F and M coprimeness are the same but Z coprimeness is a different concept. In the case of $n>2$ all these properties are distinct, but in all cases $F \subset M \subset Z$ (Johnson, Pugh and Hayton (1992), Johnson (1993), Johnson, Rogers, Pugh, Hayton, Owens (1996)). The net result is that the minimal realization problem is much more difficult in the case of $n>2$ and hence the value of the EOA is much greater (in relative terms).

We demonstrate the effectiveness of the EOA approach by the analysis of the following three examples.

Example 7.1 Consider the polynomial

$$a(s, p, z, w) = spwz - swz - pwz - sw - pw - sz + pz + sp - s + p + 1.$$

As can be easily checked, the direct application of the EOA to four-variate polynomials generally yields singular solutions. Hence, the way followed previously is:
invert the polynomial variables, *i.e.* make the substitutions $s := 1/s$, $p := 1/p$, $w := 1/w$, $z := 1/z$ to obtain

$$I(s, p, z, w) = spwz + swz - pwz + wz + sw - pw - sz - pz - s - p + 1,$$

and apply the standard EOA procedure in the form of the following sequence of operations:

```
> B1:=Augment(Ispzw);
> B1:=addcol(B1,1,2,w);
> B1:=map(simplify,addrow(B1,1,2,-z*(s*p+s-p+1)
-(s-p)));
> B2:=Augment(B1);
> B2:=addrow(B2,1,3,z);
> B2:=map(simplify,addcol(B2,1,2,s*p+s-p+1));
> B2:=map(simplify,addcol(B2,1,3,s+p));
> B3:=Augment(B2);
> B3:=addrow(B3,1,4,s);
> B3:=addcol(B3,1,4,1);
> B3:=addcol(B3,1,3,1);
> B4:=Augment(B3);
> B4:=addrow(B4,1,3,s);
> B4:=map(simplify,addcol(B4,1,4,-p-1));
> B4:=map(simplify,addcol(B4,1,5,-1));
> B4:=addrow(B4,1,3,-1);
> B4:=addrow(B4,1,5,1);
> B4:=addrow(B4,3,5,1);
> B4:=addcol(B4,2,1,-1);
> B5:=Augment(B4);
> B5:=addcol(B5,1,3,s);
> B5:=addrow(B5,1,6,-1);
> B6:=Augment (B5);
> B6:=addcol(B6,1,7,p);
> B6:=addrow(B6,1,5,-1);
> B7:=Augment (B6);
> B7:=addrow(B7,1,2,p);
> B7:=addcol(B7,1,8,-1);
> B7:=mulrow(B7,4,-1).
```

Now, apply the following row and column permutations:

```
> pr:=[1,5,3,6,2,4,7,8]:        #permutation for rows
```

```
> pc:=[2,3,5,4,1,7,8,6]: #permutation for columns
```

to obtain the singular system matrix

$$
E\Lambda - H = \begin{bmatrix}
0 & 0 & 0 & 0 & 1 & 0 & -1 & 0 \\
0 & 0 & 1 & -1 & 0 & 1 & 1 & 0 \\
\hline
0 & 1 & s & 0 & 0 & 0 & 0 & 0 \\
-1 & 0 & 0 & s-1 & 0 & 2 & 1 & 1 \\
1 & 0 & 0 & 0 & p & 0 & 0 & 0 \\
0 & 0 & 0 & -1 & 0 & p+1 & 1 & 0 \\
0 & 0 & 0 & 0 & 0 & 1 & w & 0 \\
0 & -1 & 0 & 0 & 0 & 1 & 1 & z+1
\end{bmatrix},
$$

where

$$ E\Lambda = 0I_2 \oplus sI_2 \oplus pI_2 \oplus w \oplus z $$

and

$$ E = 0I_2 \oplus I_6 . $$

Finally, apply 'reverse' variable inversion, which yields the companion matrix

$$
\Lambda - H_1 = \Lambda - EH^{-1} = \begin{bmatrix}
\sigma & 0 & 0 & 0 & 0 & 0 & 0 & 0 \\
0 & \omega & 0 & 0 & 0 & 0 & 0 & 0 \\
\hline
0 & 1 & s & 0 & 0 & -1 & 0 & 0 \\
0 & 0 & 1 & s-1 & -1 & 0 & 1 & 1 \\
1 & 0 & 1 & -1 & p-1 & 1 & 0 & 1 \\
0 & 0 & 0 & 0 & 0 & p & 1 & 0 \\
0 & 0 & 1 & -1 & -1 & 1 & w & 1 \\
0 & 0 & 0 & 1 & 1 & -1 & -1 & z
\end{bmatrix},
$$

where

$$ \Lambda = \sigma \oplus \omega \oplus sI_2 \oplus pI_2 \oplus w \oplus z $$

and σ and ω are equal to s, p, w, or z in arbitrary combinations or even can be zero. As was proved for the 2D case and can be easily repeated for the nD ($n \geq 2$) case, owing to the form of the matrix E, the first two rows of the resulting companion matrix are zero and hence they can be removed along with two first columns without changing the characteristic polynomial. Also, the particular choice of σ and ω does not influence the characteristic polynomial $\det(\Lambda - H_1)$. Hence, the final form of the companion matrix is

$$\Lambda' - H_1' = \begin{bmatrix} s & 0 & 0 & -1 & 0 & 0 \\ 0 & s-1 & -1 & 0 & 1 & 1 \\ 1 & -1 & p-1 & 1 & 0 & 1 \\ 0 & 0 & 0 & p & 1 & 0 \\ 1 & -1 & -1 & 1 & w & 1 \\ 0 & 1 & 1 & -1 & -1 & z \end{bmatrix}.$$

It is clear, however, that it is still a redundant form and that

$$\det(\Lambda' - H_1') = sp\ a(s,p,w,z).$$

In the second example we consider the generalisation of the initial polynomial representation of (3.17), which has yielded standard (nonsingular) results directly *i.e.* without any additional transformations

$$\Psi(s_1, \cdots, s_n) = \begin{bmatrix} \{s_1^{t_1}\} & \{s_1^{t_1-k_1} s_2^{k_2}\} & \cdots & \{s_1^{t_1-k_1} s_n^{k_n}\} \\ \{s_1^{k_1} s_2^{t_2-k_2}\} & \{s_2^{t_2}\} & \cdots & \{s_2^{t_2-k_2} s_n^{k_n}\} \\ \vdots & \vdots & \ddots & \vdots \\ \{s_1^{k_1} s_n^{t_n-k_n}\} & \{s_2^{k_2} s_n^{t_n-k_n}\} & \cdots & \{s_n^{t_n}\} \end{bmatrix}. \tag{7.1}$$

Example 7.2 Consider the polynomial represented as the determinant of the polynomial matrix of the form of (7.1)

$$\Psi(s,p,z,w) = \begin{bmatrix} s^2+s+1 & sp-s+1 & sz-1 & sw+w-1 \\ sp+p-1 & 2p^2+p-1 & pz+2p-1 & pw+p-1 \\ sz+2z-1 & 3pz-1 & 3z^2+z-1 & 11zw+2z-1 \\ sw-2s+1 & 7pw-1 & zw-1 & 4w^2-w+1 \end{bmatrix}$$

and apply the following chains of operations:

```
> B1:=Augment(Ispzw);

> B1:=addcol(B1,1,2,-s):

> B1:=addcol(B1,1,3,-p+1):

> B1:=addcol(B1,1,4,-z):

> B1:=addcol(B1,1,5,-w):

> B1:=map(simplify,addrow(B1,1,2,s+1));

> B1:=map(simplify,addrow(B1,1,3,p));

> B1:=map(simplify,addrow(B1,1,5,w-2));
```

```
> B2:=Augment(B1);

> B2:=addcol(B2,1,4,p+1/2);

> B2:=map(simplify,addrow(B2,1,2,1));

> B2:=map(simplify,addrow(B2,1,3,1));

> B2:=map(simplify,addrow(B2,1,4,-p-3/2));

> B2:=map(simplify,addrow(B2,1,5,-2*z));

> B2:=map(simplify,addrow(B2,1,6,-6*w-2));

> B2:=addcol(B2,1,2,1);

> B2:=addcol(B2,1,3,1);

> B2:=addcol(B2,1,6,1);

> B3:=Augment(B2);

> B3:=addcol(B3,1,6,z);

> B3:=addrow(B3,1,3,1);

> B3:=addrow(B3,1,4,1);

> B3:=map(simplify,addrow(B3,1,6,-2*z-1));

> B3:=addrow(B3,1,7,-2);

> B3:=addcol(B3,1,2,-1);

> B3:=addcol(B3,1,3,-1/2);

> B4:=Augment(B3);

> B4:=addcol(B4,1,8,w);

> B4:=map(simplify,addrow(B4,1,4,1));

> B4:=map(simplify,addrow(B4,1,7,-10*z));

> B4:=map(simplify,addrow(B4,1,8,-3*w+5));

> B4:=addcol(B4,2,1,-5);

> B4:=addcol(B4,1,3,-2);

> B4:=addcol(B4,1,4,-5/3);

> B4:=addcol(B4,1,5,-2);

> B4:=addcol(B4,1,6,-2/3);

> B4:=addcol(B4,3,7,2);

> B4:=mulrow(B4,4,-1);

> B4:=mulrow(B4,6,-1);
```

```
> B4:=mulrow(B4,7,-1/2);
```

```
> B4:=mulrow(B4,8,-1/3).
```

Now, apply the following row and column permutations:

```
> pr:=[4,5,3,6,2,7,1,8]:        #permutation for rows
```

```
> pc:=[5,4,6,3,7,2,8,1]: #permutation for columns
```

to yield directly the nonsingular companion matrix

$$
\begin{bmatrix}
s-9 & -6/49 & -25/6 & -8 & -16 & -1 & -1 & 4 \\
12 & s+59/6 & 35/6 & 10 & 19 & 1 & 0 & -5 \\
1 & 1 & p+1/2 & 1 & 2 & 0 & 1 & 0 \\
5/2 & 3/2 & 7/4 & p+3/2 & 4 & 0 & 5/2 & 0 \\
10 & 47/6 & 10/3 & 9 & z+18 & 1 & 0 & -5 \\
11/2 & 47/12 & 13/6 & 9/2 & 19/2 & z+1/2 & 1/2 & -5/2 \\
-2 & -5/3 & -2/3 & -2 & -4 & 0 & w & 1 \\
31/3 & 28/3 & 14/3 & 10 & 61/3 & 2/3 & 1/3 & w-5
\end{bmatrix}.
$$

Note here that direct application of the EOA (*i.e.* no variable transformations have been applied) has resulted in a standard (nonsingular) realization with the absolute minimum dimension.

The similar results can be established when applying the initial representation of the form (3.18). The other, still open question is whether and when such representations exist for a given n-variate polynomial.

In what follows, we demonstrate the potential of the EOA in the $n>2$ case.

Example 7.3 In this example the following polynomial matrix (*c.f.* Cockburn 1998)

$$
F(x,y,z) = \begin{bmatrix}
0 & -2yz & 2y^2 & 4y^2-4z^2 & -4xy & 4xz \\
2yz & 0 & -2xy & -3xy & 3x^2-3z^2 & 3yz \\
-2y & 2xy & 0 & xz & -yz & y^2-x^2
\end{bmatrix}
$$

is taken to represent in the associated system matrix form. The EOA here is to apply the following chain of operations

```
> with(linalg):
```

```
> Fxyz:=matrix(3,6,[0,-2*y*z,2*y^2,4*y^2-4*z^2,-
    4*x*y,4*x*z, 2*y*z,0,-2*x*y,-3*x*y,3*x^2-
    3*z^2,3*y*z,-2*y,2*x*y,0,x*z,-y*z,y^2-x^2]);
```

```
> B1:=Augment(Fxyz);
```

```
> B1:=addcol(B1,1,2,y);
```

```
> B1:=map(simplify,addrow(B1,1,3,-2*(z)));
```

```
> B1:=map(simplify,addrow(B1,1,4,2));

> B2:=Augment(B1);

> B2:=addcol(B2,1,4,y);

> B2:=addrow(B2,1,3,2*(z));

> B2:=addrow(B2,1,5,-2*(x));

> B3:=Augment(B2);

> B3:=addcol(B3,1,6,y);

> B3:=addrow(B3,1,4,-2*(y));

> B3:=addrow(B3,1,5,2*(x));

> B4:=Augment(B3);

> B4:=addrow(B4,1,5,y);

> B4:=addcol(B4,1,2,2);

> B4:=map(simplify,addcol(B4,1,8,-4*(y)));

> B4:=map(simplify,addcol(B4,1,9,4*(x)));

> B5:=Augment(B4);

> B5:=addrow(B5,1,7,x);

> B5:=addcol(B5,1,3,-2);

> B5:=addcol(B5,1,9,3/y);

> B5:=map(simplify,addcol(B5,1,10,-3/x));

> B6:=Augment(B5);

> B6:=addrow(B6,1,9,x);

> B6:=addcol(B6,1,5,2);

> B6:=addcol(B6,1,10,-z);

> B6:=map(simplify,addcol(B6,1,12,x));

> B6:=addrow(B6,2,3,4/3);

> B7:=Augment(B6);

> B7:=addrow(B7,1,8,z);

> B7:=addcol(B7,1,6,-2);

> B7:=map(simplify,addcol(B7,1,11,4/z));

> B7:=map(simplify,addcol(B7,1,13,-4/x));

> B7:=map(simplify,addrow(B7,2,1,4));
```

```
> B8:=Augment(B7);

> B8:=addrow(B8,1,10,z);

> B8:=addcol(B8,1,8,2);

> B8:=map(simplify,addcol(B8,1,13,3/z));

> B8:=map(simplify,addcol(B8,1,14,-3/y));

> B9:=Augment(B8);

> B9:=addrow(B9,1,12,y);

> B9:=addcol(B9,1,14,z);

> B9:=addcol(B9,1,15,-y);

> B9:=addrow(B9,1,2,-3);

> B10:=Augment(B9);

> B10:=addcol(B10,1,2,y);

> B10:=addrow(B10,1,13,-1);

> B11:=Augment(B10);

> B11:=addcol(B11,1,4,z);

> B11:=addrow(B11,1,13,-1);

> B12:=Augment(B11);

> B12:=addcol(B12,1,6,z);

> B12:=addrow(B12,1,13,-1);

> B13:=Augment(B12);

> B14:=Augment(B13);

> B15:=Augment(B14);

> B15:=addcol(B15,1,21,y);

> B15:=addrow(B15,1,7,1):

> B15:=addcol(B15,2,20,x):

> B15:=addrow(B15,2,11,3):

> B15:=addcol(B15,3,19,z):

> B15:=addrow(B15,3,10,1);

> B16:=Augment(B15):

> B17:=Augment(B16):

> B18:=Augment(B17);
```

```
> B18:=addrow(B18,1,14,y):

> B18:=addcol(B18,1,22,-3):

> B18:=addrow(B18,2,4,y):

> B18:=addcol(B18,2,24,-1):

> B18:=addrow(B18,3,5,x):

> B18:=addcol(B18,3,23,-1);

> A:=B18.
```

Now apply the row and column permutations

```
pr:=[15,12,11,5,13,20,21,4,9,14,16,17,18,19,6,7,8,10,1
    ,2,3]:    #permutation for rows

pc:=[16,17,18,3,24,14,13,2,10,1,21,20,19,15,22,12,11,2
    3,4,5,6,7,8,9]: #permutation for columns
```

which produce the matrix AA and then apply the following operations

```
> F:=mulrow(AA,1,-3/2):

> F:=mulrow(F,2,1/6):

> F:=mulrow(F,3,1/2);

> F:=addcol(F,1,6,2):

> F:=addcol(F,1,14,3/2):

> F:=addcol(F,2,7,-2/3):

> F:=addcol(F,2,16,-1/6);

> F:=addcol(F,3,9,3/2):

> F:=addcol(F,3,17,-1/2);

> FF:=submatrix(F,4..21,4..24);
```

to yield the final system matrix in the form

$$xI_4 \oplus yI_7 \oplus zI_4 \oplus 0_{3,6} - H \triangleq$$

$$=
\left[
\begin{array}{cccc|ccccccc|cccc|cccccc}
x & 0 & 0 & 0 & 0 & 0 & 0 & 0 & 0 & 0 & 0 & 0 & 0 & 0 & 0 & 0 & 1 & 0 & 0 & 0 & 0 \\
0 & x & 0 & -1/3 & 0 & 0 & 0 & 0 & 0 & 0 & 0 & 0 & -1/3 & 0 & 0 & 0 & 0 & 1 & 0 & 0 & 0 \\
0 & 0 & x & 0 & 0 & 0 & 0 & 0 & 0 & 0 & 0 & 0 & 0 & 0 & 0 & 0 & 0 & 0 & 0 & -1 & 0 \\
0 & 0 & 0 & x & 0 & 3 & 0 & 0 & 0 & 0 & 0 & 0 & 0 & -1 & 0 & 0 & 0 & 0 & 0 & 0 & -1 \\ \hline
0 & 0 & 0 & 0 & y & 0 & 0 & 0 & 0 & 0 & 0 & 0 & 0 & 0 & 0 & 1 & 0 & 0 & 0 & 0 & 0 \\
0 & 0 & 0 & 0 & 0 & y & 0 & 0 & 0 & 0 & 0 & 0 & 0 & 0 & 0 & 0 & 0 & 0 & 0 & 0 & 1 \\
0 & 0 & -3 & 0 & 0 & 0 & y & 0 & 0 & 0 & -3 & 0 & 0 & 0 & 0 & 0 & 3 & 0 & 0 & 0 & 0 \\
0 & 0 & 2 & 0 & 0 & 0 & 0 & y & 0 & 0 & 3/2 & 0 & 0 & 0 & 0 & 0 & 0 & 0 & 0 & 0 & 0 \\
0 & 0 & 0 & -2/3 & 0 & 0 & 0 & 0 & y & 0 & 0 & 0 & -1/6 & 0 & 0 & 0 & 0 & 0 & 0 & 0 & 0 \\
0 & 0 & 0 & 0 & 0 & 3/2 & 0 & 0 & 0 & y & 0 & 0 & 0 & -1/2 & 0 & 0 & 0 & 0 & 0 & 0 & 0 \\
0 & 0 & 0 & 0 & 0 & 0 & 0 & 0 & 0 & 0 & y & 0 & 0 & 0 & 0 & 0 & 0 & 0 & -1 & 0 & 0 \\ \hline
0 & 0 & 0 & 0 & 0 & 0 & 0 & 0 & 0 & 0 & 0 & z & 0 & 0 & 0 & 0 & 0 & 1 & 0 & 0 & 0 \\
0 & 0 & 0 & 0 & 0 & 0 & 0 & 0 & 0 & 0 & 0 & 0 & z & 0 & 0 & 0 & 0 & 0 & 1 & 0 & 0 \\
0 & 0 & 0 & 0 & 0 & 0 & 0 & 0 & 0 & 0 & 0 & 0 & 0 & z & 0 & 0 & 0 & 0 & 0 & 1 & 0 \\
0 & 0 & 0 & 0 & 0 & 0 & 0 & 0 & 0 & 0 & 0 & 0 & 0 & 0 & z & 1 & 0 & 0 & 0 & 0 & 0 \\ \hline
0 & 0 & 0 & 0 & 0 & 0 & 1 & 0 & 0 & 0 & 0 & -3 & 0 & 0 & 0 & 0 & 0 & 0 & 0 & 0 & 0 \\
0 & -1 & 0 & 0 & 1 & 0 & 0 & 0 & 0 & 0 & 0 & 0 & 0 & 0 & 0 & 0 & 0 & 0 & 0 & 0 & 0 \\
1 & 0 & 0 & 0 & 0 & 0 & 0 & 0 & 0 & 0 & 0 & 0 & 0 & 0 & -1 & 0 & 0 & 0 & 0 & 0 & 0 \\
\end{array}
\right]$$

Here, the matrix H is clearly has the block form

$$H = \begin{bmatrix} H_{11} & H_{12} \\ H_{21} & H_{22} \end{bmatrix}, \tag{7.2}$$

where H_{11}, H_{12}, H_{21} and H_{22} are blocks of the dimensions 15×15, 15×6, 3×15 and 3×6, respectively. It is easy to check that

$$F(x, y, z) = H_{22} + H_{21}[\Lambda - H_{11}]^{-1} H_{12}, \tag{7.3}$$

where

$$\Lambda = x^{-1}I_4 \oplus y^{-1}I_7 \oplus z^{-1}I_4. \tag{7.4}$$

In this case, the EOA has again directly provided (without any preliminaries) the solution to the problem of how to represent a 3D polynomial matrix in the form of the 3D system matrix of the Roesser form. For applications to this problem see Cockburn, Morton (1997).

7.2 Application to Repetitive Processes

As discussed in Chapter 1, repetitive processes are a class of 2D linear systems whose unique characteristic is a series of sweeps, termed passes, through a set of dynamics defined over a fixed and finite duration known as the pass length. On each pass an output, termed the pass profile, is produced which acts as a forcing function on, and hence contributes to, the dynamics of the next pass profile. Industrial examples include long-wall coal cutting and metal rolling operations, *c.f.* Rogers, Owens (1992), Smyth (1992). Algorithmic examples include certain classes of iterative learning control schemes (Amann, Owens and Rogers (1998)) and iterative solution algorithms for classes of nonlinear dynamic optimal control problems based on the maximum principle (Roberts (2000)). Both of these algorithmic examples are based on the sub-class of so-called discrete linear repetitive processes, which are the subject of this section.

In effect, the 2D systems structure of repetitive processes arises from the need to use two co-ordinates to specify a variable, *i.e.* the pass number or index $k \geq 0$ and the position t along the finite pass length. A key difference with, in particular, 2D discrete linear systems described by the well known Roesser or Fornasini-Marchesini state space models arises from the finite pass length, *i.e.* the duration of information propagation in one of the two separate directions.

The basic unique control problem for repetitive processes arises directly from the interaction between successive pass profiles and can result in the output sequence of pass profiles containing oscillations that increase in amplitude in the pass to pass direction. Such behaviour is easily generated in simulation studies and in experiments on scaled models of industrial examples and, in particular, the long-wall coal cutting process (Rogers, Owens (1992)). Also, this problem cannot be removed by direct application of existing control techniques such as standard linear time invariant, or 1D, feedback control schemes. In effect, the reason for this is that such an approach essentially ignores their finite pass length repeatable nature and the effects of resetting the initial conditions before the start of each pass.

Motivated by these control difficulties, a rigorous stability theory for linear constant pass length repetitive processes has been developed (Rogers, Owens (1992)). This theory is based on an abstract model of the underlying dynamics formulated in a Banach space setting, which includes all such processes as special cases. Physically the strongest version of this theory can be interpreted (in terms of the norm on the underlying function space) as a form of bounded-input/bounded-output stability independent of the pass length and is known as stability along the pass. Gałkowski, Rogers and Owens (1996), (1998) have used their 2D systems interpretations (in the Roesser and Fornasini-Marchesini forms) to provide a complete characterization of some key controllability properties for these processes.

The state-space model of the sub-class of discrete linear repetitive processes has the structure

$$X(k+1,t+1) = AX(k+1,t) + BU(k+1,t) + B_0\widetilde{Y}(k,t),$$

$$\widetilde{Y}(k+1,t) = CX(k+1,t) + DU(k+1,t) + D_0\widetilde{Y}(k,t). \tag{7.5}$$

Here on pass k, $X(k, t)$ is the $n \times 1$ state vector at sample t along the pass length $\alpha < +\infty$, $\widetilde{Y}(k,t)$ is the $m \times 1$ vector pass profile, and $U(k, t)$ is the $p \times 1$ vector of control inputs.

This model is very similar to the standard Roesser form with the exception that information propagation in one of the two separate directions (along the pass) only occurs over a finite duration (the pass length) and the process output equation has the form

$$Y(k,t) = \widetilde{Y}(k,t) \tag{7.6}$$

or

$$Y(k,t) = \begin{bmatrix} 0 & I \end{bmatrix} \begin{bmatrix} X(k,t) \\ \widetilde{Y}(k,t) \end{bmatrix} + 0U(k,t). \tag{7.7}$$

The equations of (7.5) can be transformed to the standard Roesser state-space model by applying a simple forward transformation of the pass profile vector followed by a change of variable in the pass number (Gałkowski, Rogers, Gramacki A., Gramacki J. and Owens (1999)) but the feature of (7.7) is still valid.

The EOA approach can clearly be applied to the repetitive processes described by their transfer function matrices (Rogers, Owens 1992) and, gives some new insights into the equivalence properties of this system class.

Examine hence the following example.

Example 7.4 Consider the discrete linear repetitive process described by the matrices

$$A = \begin{bmatrix} 1 & -1 \\ 2 & 0 \end{bmatrix}, \quad B_0 = \begin{bmatrix} 0 & 1 \\ 2 & 1 \end{bmatrix}, \quad B = \begin{bmatrix} 1 & 0 \\ 1 & 1 \end{bmatrix},$$

$$C = \begin{bmatrix} 0 & 0 \\ 0 & 1 \end{bmatrix}, \quad D = \begin{bmatrix} 1 & 0 \\ 2 & -1 \end{bmatrix}, \quad D_0 = \begin{bmatrix} -1 & 0 \\ 0 & 0 \end{bmatrix}.$$

The linear repetitive processes can be characterised by the transfer function matrix (*c.f.* Rogers, Owens 1992) of the form

$$G(s, p) = s\left(sI_m - D_0 - C(pI_n - A)^{-1}B_0\right)^{-1}\left(D + C(pI_n - A)^{-1}B\right), \tag{7.8}$$

where 's' and 'p' represent the complex, generalised frequencies for the information propagation in the 'from pass to pass' and 'along the pass' directions, respectively. In this particular case, the transfer function matrix of (7.8) is equal to

$$G(s,p) = \begin{bmatrix} \dfrac{s}{s+1} & 0 \\ \dfrac{(sp+s+3p-1)s}{(s+1)(sp^2-sp+2s-p-1)} & \dfrac{(p^2-2p+3)s}{(sp^2-sp+2s-p-1)} \end{bmatrix}.$$

As direct application of the EOA would almost certainly yield the singular solution, first apply the variable substitution $p := 1/p$, $s := 1/s$ which leads to the following transfer function matrix

$$H(s,p) = \begin{bmatrix} \dfrac{1}{s+1} & 0 \\ \dfrac{(sp-3s-p-1)p}{(s+1)(sp^2+sp-2p^2+p-1)} & \dfrac{(3p^2-2p+1)}{(sp^2+sp-2p^2+p-1)} \end{bmatrix}.$$

The initial polynomial matrix representation for the EOA is, for example

$$H_0(s,p,z) = \begin{bmatrix} z(s+1)(sp^2+sp-2p^2+p-1) & 0 \\ -(sp^2+sp-2p^2+p-1) & \\ \hline -(sp-3s-p-1)p & z(sp^2+sp-2p^2+p-1) \\ & -(3p^2-2p+1) \end{bmatrix}.$$

Now, apply the following EOA procedure

```
> H1:=Augment(H0):
> H1:=addrow(H1,1,2,z):
> H1:=addcol(H1,1,2,-(s+1)*(s*p^2-2*p^2+s*p+p-1));
> H2:=Augment(H1):
> H2:=addrow(H2,1,4,z):
> H2:=map(simplify,addcol(H2,1,4,-(s*p^2-2*p^2+s*p+p-
  1)));
> H3:=Augment(H2):
> H3:=addcol(H3,1,4,(s*p^2-2*p^2+s*p+p-1)):
> H3:=addrow(H3,1,3,s+1):
> H3:=addrow(H3,1,4,1);
> H4:=Augment(H3):
> H4:=addcol(H4,1,5,p);
> H4:=map(simplify,addrow(H4,1,2,-s*p-s+2*p-1));
> H4:=map(simplify,addrow(H4,1,6,s*p-3*s-p-1));
> H5:=Augment(H4):
```

```
> H5:=addcol(H5,1,2,s):
> H5:=map(simplify,addrow(H5,1,7,-p+3)):
> H5:=map(simplify,addrow(H5,1,3,p+1)):
> H5:=addrow(H5,3,7,1):
> H5:=addcol(H5,1,2,-2);
> H6:=Augment(H5):
> H6:=addcol(H6,1,8,p):
> H6:=map(simplify,addrow(H6,1,5,s*p-2*p+s+1));
> H6:=map(simplify,addrow(H6,1,8,3*p-2));
> H7:=Augment(H6):
> H7:=addrow(H7,1,6,s-2):
> H7:=map(simplify,addcol(H7,1,2,-p-1)):
> H7:=addrow(H7,1,9,3);
> H8:=Augment(H7):
> H8:=addcol(H8,1,5,p):
> H8:=map(simplify,addrow(H8,1,10,-1));
> H9:=Augment(H8):
> H9:=addrow(H9,1,5,s):
> H9:=addcol(H9,1,6,-1):
> H9:=mulrow(H9,3,-1).
```

Now, apply the following row and column permutations:

```
> pr:=[1,5,8,9,2,3,4,6,7,10,11]:      #permutation for
  rows
> pc:=[2,1,3,7,6,4,11,10,5,9,8]: #permutation for
  columns
```

to yield the following singular system matrix

$$\begin{bmatrix}
0 & 1 & 0 & 0 & -1 & 0 & 0 & 0 & 0 & 0 & 0 \\
0 & s & 0 & 0 & -2 & 0 & 0 & 0 & 1 & 0 & 0 \\
0 & 0 & s-2 & 0 & 0 & 3 & 1 & 0 & 0 & 0 & 1 \\
0 & 0 & 0 & s+1 & 0 & 0 & 0 & 0 & 0 & 1 & 0 \\
1 & 0 & 0 & 0 & p & 0 & 0 & 0 & 0 & 0 & 0 \\
0 & 0 & -1 & 0 & 0 & p+1 & 0 & 0 & 0 & 0 & 0 \\
0 & 0 & 0 & 0 & 0 & 1 & p & 0 & 0 & 0 & 0 \\
0 & 0 & 0 & 0 & 1 & 0 & 0 & p & 0 & 0 & 0 \\
0 & 0 & 0 & 1 & -3 & 0 & 0 & -1 & p+1 & 0 & 0 \\
0 & 0 & 0 & 1 & 0 & 0 & 0 & 0 & 0 & z & 0 \\
-1 & 0 & 3 & 1 & -10 & -5 & -1 & -1 & 4 & 0 & z
\end{bmatrix}.$$

Finally, apply reverse variable inversion and remove the first column and row, *i.e.* apply the operations

```
> X:=diag(0,s$3,p$5,z$2):
> AAA:=evalm(X-AA);
> A1:=submatrix(AAA,1..9,1..9);
> B1:=submatrix(AAA,1..9,10..11);
> C1:=submatrix(AAA,10..11,1..9);
> D1:=submatrix(AAA,10..11,10..11);
> E:=diag(0,1$8);
> Ax:=evalm(inverse(A1)&*E);
> Bx:=evalm(-inverse(A1)&*B1);
> Cx:=evalm(C1&*Ax);
> Dx:=evalm(D1-Cx&*B);
> Ax:=evalm(inverse(A1)&*E);
> Ay:=submatrix(Ax,2..9,2..9);
> By:=submatrix(Bx,2..9,1..2);
> Cy:=submatrix(Cx,1..2,2..9);
> Dy:=Dx;
```

to obtain the state space realization of the Roesser type with the following model matrices

$$A = A_y = \begin{bmatrix} 0 & 0 & 0 & 0 & 0 & 0 & -1 & 0 \\ 0 & 0 & 0 & 0 & 1 & -1 & 0 & 0 \\ 0 & 0 & -1 & 0 & 0 & 0 & 0 & 0 \\ 0 & 0 & 0 & 0 & 0 & 0 & -1 & 0 \\ 0 & 0 & 0 & 0 & 0 & -1 & 0 & 0 \\ 0 & -1 & 0 & 0 & 2 & 1 & 0 & 0 \\ -1 & 0 & -1 & 0 & 0 & 0 & 1 & 1 \\ -1 & 0 & 0 & 0 & 0 & 0 & -2 & 0 \end{bmatrix}, \quad B = B_y = \begin{bmatrix} 0 & 0 \\ 0 & 0 \\ 0 & 0 \\ -1 & 0 \\ 0 & 0 \\ 0 & 0 \\ 0 & -1 \\ -1 & 0 \end{bmatrix},$$

$$C = C_y = \begin{bmatrix} 0 & 0 & 1 & 0 & 0 & 0 & 0 & 0 \\ 3 & -1 & 0 & -1 & -1 & -1 & -1 & 1 \end{bmatrix}, \quad D = D_y = \begin{bmatrix} 1 & 0 \\ 0 & -1 \end{bmatrix}.$$

It is easy to check that the transfer function matrix related to this state space realization, calculated by the Maple subroutine as

```
> FF:=evalm(Dy+Cy&*inverse(diag(s$3,p$5)-Ay)&*By):
> map(simplify,FF);
```

is the original transfer function matrix. However, note that one of the essential features of the repetitive processes of the form of (7.7) has been lost. In particular, the repetitive process has been embedded into an equivalent 2D standard form of the Roesser model without the feature of (7.7) being (explicitly) present. Also, this realisation is not minimal and the question of what conditions are required for this property is still open.

7.3 The Elementary Operation Algorithm and Laurent Polynomials

The Laurent polynomials have found an extensive use in polynomial matrix based analysis of both 1D and nD linear systems. Such a polynomial has entries with both positive and negative powers of indeterminates present, and the use of such polynomials clearly implies some form of singularity. In the discrete case, it arises because the variable s_i denotes a forward (in the time or space) shift and s_i^{-1} denotes a backward shift.

Hence, non-causality is the intrinsic feature of such systems, since singularity has clear links to causality (Gałkowski (2000b), Gregor (1993), Kaczorek (1992a), Lewis (1992), Uetake (1992)). Causality is a key feature of classical dynamical systems (it must be present for physical realizability). In the 1D case, the general concept of "time" imposes a natural ordering into "past", "present" and "future". Again, the situation is different in the nD case due, in effect, to the fact that some of the indeterminates have a spatial rather than a temporal characteristic. Hence, causality is not required since it is only necessary to be able to recursively apply the required computations. Singular systems are commonly described by the so-called descriptor (singular state-space) equations instead of standard state-space equations, where the shifted vector on the left hand side is left multiplied by a rectangular or square, singular matrix, *c.f.* Preliminaries (2.5), (2.6). An important point to note is that sometimes it is possible to avoid using singular models by applying well defined algebraic operations (Gałkowski (1992a)). For example, it is sometimes as simple as reformulating the state vector or changing the model type. The detailed treatment of this problem is provided in (Gałkowski (2000b)).

The Laurent polynomials turn out to be a very promising way of describing such systems (Fornasini (1999)), Jeżek (1983), (1989), Wood, Rogers, Owens (1998)). In this section we show an example of using the EOA in combination with the Laurent polynomials approach, which yields an interesting result giving some new promising directions to future work.

Example 7.5 Consider the proper, but non-principal, transfer function matrix

$$G(s, p) = \begin{bmatrix} \dfrac{s}{s+p-1} \\ \dfrac{p}{s-p-1)} \end{bmatrix}. \tag{7.9}$$

Note here the fact that $ss^{-1} = 1$, $pp^{-1} = 1$ and rewrite the transfer function matrix $G(s,p)$ in the following alternative form

$$G_1(s_1, s_2, p_1, p_2) = \begin{bmatrix} \dfrac{s_2\, p_1\, p_2}{1 - s_2\, p_1\, p_2 + s_1 s_2 p_2 + 2s_2 - 2s_1 s_2\, p_1\, p_2} \\ \dfrac{s_1 s_2 p_2}{1 + s_2\, p_1\, p_2 - s_2\, s_1\, p_2 - 2s_1 s_2\, p_1\, p_2} \end{bmatrix} \tag{7.10}$$

where $s_1 := s^{-1}, s_2 := s$, $p_1 = p^{-1}, p_2 := p$. Note that, in fact there are infinitely many such four-variate representations. Now, introduce the polynomial matrix associated to (7.10)

$$B(s_1, s_2, p_1, p_2) = \begin{bmatrix} z_1\big(1 - s_2\, p_1\, p_2 + s_1 s_2 p_2 + 2s_2 - 2s_1 s_2\, p_1\, p_2\big) - s_2\, p_1\, p_2 \\ z_2\big(1 + s_2\, p_1\, p_2 - s_2\, s_1\, p_2 - 2s_1 s_2\, p_1\, p_2\big) - s_1 s_2 p_2 \end{bmatrix}$$

and apply the following EOA operation chain:

```
> B1:=Augment(B);
> B1:=addcol(B1,1,2,z1);
> B1:=map(simplify,addrow(B1,1,2,-(1+2*s2-
  s2*p1*p2+s1*s2*p2-2*s1*s2*p1*p2)));
> B2:=Augment(B1);
```

```
> B2:=addcol(B2,1,3,z2);

> B2:=map(simplify,addrow(B2,1,4,-(1+s2*p1*p2-
  s1*s2*p2-2*s1*s2*p1*p2)));

> B3:=Augment(B2);

> B3:=addrow(B3,1,4,s2);

> B3:=map(simplify,addcol(B3,1,3,-2*p1*p2*s1+s1*p2-
  p1*p2+2));

> B3:=addcol(B3,1,4,p1*p2);

> B4:=Augment(B3);

> B4:=addrow(B4,1,6,s2);

> B4:=map(simplify,addcol(B4,1,3,p1*p2-s1*p2-
  2*s1*p1*p2));

> B4:=addcol(B4,1,5,s1*p2);

> B5:=Augment(B4);

> B5:=addrow(B5,1,2,p2);

> B5:=map(simplify,addcol(B5,1,4,-p1+s1+2*s1*p1));

> B5:=addcol(B5,1,6,-s1);

> B6:=Augment(B5);

> B6:=addrow(B6,1,4,p2);

> B6:=map(simplify,addcol(B6,1,6,2*s1*p1+p1-s1));

> B6:=map(simplify,addcol(B6,1,7,-p1));

> B7:=Augment(B6);

> B7:=addrow(B7,1,2,p1);

> B7:=map(simplify,addcol(B7,1,7,-2*s1-1));

> B7:=map(simplify,addcol(B7,1,8,1));

> B7:=addrow(B7,1,2,-1/2);

> B7:=mulrow(B7,1,-1/2);

> B8:=Augment(B7);

> B8:=addrow(B8,1,4,s1);

> B8:=map(simplify,addcol(B8,1,7,-2*p1-1));

> B8:=map(simplify,addcol(B8,1,9,1));
```

```
> B8:=addrow(B8,1,4,-1/2);

> B8:=mulrow(B8,1,-1/2).
```

Now, apply the following row and column permutations:

```
> pr:=[2,4,9,10,1,3,5,6,8,7]:        #permutation for
  rows

> pc:=[8,1,6,5,7,2,4,3,9]: #permutation for columns
```

to obtain the system matrix

$$
\begin{bmatrix}
s_1+1/2 & 0 & 0 & 0 & 0 & -1/2 & 0 & 0 & -1/2 \\
0 & s_1-1/2 & 0 & 0 & 1/2 & 0 & 1 & 0 & -1/2 \\
-1 & 0 & s_2 & 0 & 0 & 0 & 0 & 0 & 0 \\
0 & 0 & 0 & s_2 & -1 & 0 & 0 & 0 & 0 \\
0 & -1/2 & 0 & 0 & p_1+1/2 & 0 & 0 & 0 & -1/2 \\
1/2 & 0 & 0 & 0 & 0 & p_1-1/2 & 0 & 1 & -1/2 \\
0 & 0 & 0 & 1 & 0 & 0 & p_2 & 0 & 0 \\
2 & 0 & 1 & 0 & 0 & 0 & 0 & p_2 & 0 \\
1 & 0 & 0 & 0 & 0 & 0 & 0 & 0 & z_1 \\
0 & 0 & 0 & 0 & 1 & 0 & 0 & 0 & z_2
\end{bmatrix}.
$$

Note that applying here the Laurent polynomials technique allows us to obtain standard (nonsingular) state-space realizations for intrinsically singular (non-principal) systems. However, physical realization requirements allow us to invert variables, which refers obviously to the inverted direction of the information propagation, only for the space directions, but not to the time direction. Hence, in practical applications at least one variable, which represents the time evolution, should not be inverted. This is in contrast to Example 7.5, which however shows only potential possibilities and does not aim to be a full study of the problem.

8. State-Space Realizations for 2D/nD Systems Revisited – Relations to the EOA

8.1 Overview

One of the first major contributions to the problem of constructing state space realizations for SISO 2D discrete linear systems described by a two-variate transfer function was given in a two-part paper by Kung *et al* (1977). It is interesting to note here that a partial result of this form was obtained independently by Gałkowski and Piekarski (1978). Hence, Kung *et al* have proposed an *a priori* non-minimal realization for the rational transfer function

$$H(s_1^{-1}, s_2^{-1}) = \frac{b(s_1^{-1}, s_2^{-1})}{a(s_1^{-1}, s_2^{-1})} = \frac{\displaystyle\sum_{i=0}^{n}\sum_{j=0}^{m} b_{ij} s_1^{-1} s_2^{-1}}{\displaystyle\sum_{i=0}^{n}\sum_{j=0}^{m} a_{ij} s_1^{-1} s_2^{-1}}, \tag{8.1}$$

of the form

$$\begin{matrix} m\{ \\ n\{ \\ n\{ \end{matrix} \begin{bmatrix} x_h(i+1, j) \\ x_{v_1}(i, j+1) \\ x_{v_2}(i, j+1) \end{bmatrix} = A \begin{bmatrix} x_h(i, j) \\ x_{v_1}(i, j) \\ x_{v_2}(i, j) \end{bmatrix} + bu(i, j), \tag{8.2}$$

$$y(i, j) = cx(i, j), \tag{8.3}$$

where

$$A = \begin{bmatrix} -a_{10} & \cdots & -a_{n-1,0} & -a_{n0} & -1 & & & & \\ 1 & & & 0 & & & & & \\ & \ddots & 0 & & & 0 & & 0 & \\ 0 & & 1 & 0 & & & & & \\ \hline \tilde{a}_{11} & \cdots & \cdots & \tilde{a}_{n1} & -a_{01} & 1 & & 0 & \\ \vdots & \ddots & & \vdots & & 0 & \ddots & & \\ \vdots & & \ddots & \vdots & & & 1 & & 0 \\ \tilde{a}_{1m} & \cdots & \cdots & \tilde{a}_{nm} & -a_{0m} & & 0 & & \\ \hline \tilde{b}_{11} & \cdots & \cdots & \tilde{b}_{n1} & -b_{01} & & & 0 & 1 \\ \vdots & \ddots & & \vdots & & & & & \ddots \\ \vdots & & \ddots & \vdots & & 0 & & & 1 \\ \tilde{b}_{1m} & \cdots & \cdots & \tilde{b}_{nm} & -b_{0m} & & & 0 & \end{bmatrix},$$

$$b = \begin{bmatrix} 1 \\ 0 \\ \vdots \\ 0 \\ \hline a_{01} \\ \vdots \\ a_{0m} \\ \hline b_{01} \\ \vdots \\ b_{0m} \end{bmatrix}, \quad c = \begin{bmatrix} \tilde{b}_{10} & \cdots & \tilde{b}_{n0} & -b_{00} & 0 & \cdots & 0 & 1 & 0 & \cdots & 0 \end{bmatrix}, \qquad (8.4)$$

$$\tilde{a}_{ij} = a_{ij} - a_{i0}a_{0j}, \quad \tilde{b}_{ij} = b_{ij} - b_{i0}b_{0j}, \quad 1 \le i \le n, 1 \le j \le m. \qquad (8.5)$$

This result is extremely significant since, although it only deals with the 2D SISO case, it provided the essential starting point for much further work - including the following:

a. Kurek (1993) for 2D SISO systems,

b. Badreddin and Mansour (1983), (1984) for 2D MIMO systems with extensions by Kummert (1989) and Premaratne, Jury and Mansour (1990),

c. Theodoru and Tzafestas (1984a), Manikopoulos and Antoniou (1990), Paraskevopoulos and Kiritsis (1993) for 3D systems, and

d. Theodorou and Tzafestas (1984b) and Venkateswarlu, Eswaran and Antoniu (1990) for the general nD systems case.

Other work in this area by Gałkowski (1981a) will be considered in the next section.

Kung *et al* have also provided essential insights into the problem of existence and construction of the minimal realization for SISO 2D systems. They showed that using block diagonal similarity transforms yields the so-called modal controller form $\{A, B, C\}$ (D is assumed zero)

$$
A = \left[\begin{array}{ccccc|c}
-a_{10} & \cdots & -a_{n-1,0} & -a_{n0} & & \\
1 & & & 0 & & \\
& \ddots & & & & A_{12} \\
0 & & 1 & 0 & & \\
\hline
& & & & -a_{01} & \cdots & -a_{0,m-1} & -a_{0,m} \\
& & & & 1 & & & \\
& A_{21} & & & & \ddots & & \\
& & & & 0 & & 1 & 0
\end{array}\right],
$$

$$
B = \left[\begin{array}{c}
1 \\ 0 \\ \vdots \\ 0 \\ \hline 1 \\ 0 \\ \vdots \\ 0
\end{array}\right]\begin{array}{l} \left.\right\}n \\ \\ \left.\right\}m \end{array}, \quad C = \begin{bmatrix} b_{10} & \cdots & b_{n0} & | & b_{01} & \cdots & b_{0m} \end{bmatrix}, \tag{8.6}
$$

and A_{12} and A_{21} should be chosen to satisfy

$$
\det\left[\begin{pmatrix} s_1 I_n & 0 \\ 0 & s_2 I_m \end{pmatrix} - A\right] = a(s_1, s_2) \tag{8.7}
$$

and

$$
\det\left[\begin{array}{c|c} \begin{pmatrix} s_1 I_n & 0 \\ 0 & s_2 I_m \end{pmatrix} - A & B \\ \hline -C & 0 \end{array}\right] = b(s_1, s_2). \tag{8.8}
$$

Note that in the previous analysis the inverse variables s_1^{-1} and s_2^{-1} are used instead of s_1 and s_2. This is mathematically equivalent on multiplying/dividing both the transfer function numerator and denominator polynomials by the term $s_1{}^n s_2{}^m$, where n and m are maximal degrees of the polynomial concerned. As mentioned in Section 7.3, the inverted variables are frequently used for discrete systems to stress the practical applicability of results. Namely, it is obvious that for the discrete systems the variable $s_i, i = 1,2$ has the meaning of forward shifting the signal along the direction i, which is, in fact, a non-causal, physically non-realizable operation, when the variable is related to the time. For 2D/nD systems it is frequently admissible but only when related to the space direction. However, in many applications, such as digital filtering it is preferred to deal with completely causal systems and this is the reason that we use inverted variables $s_i^{-1}, i = 1,2$ since they represent the delay operation which is always causal and hence physically realizable.

It is straightforward to see that, to obtain the minimal realization from (8.6)-(8.8) one has to solve $2(m+1)(n+1)$ nonlinear – algebraic equations with $2mn$ variables (the entries of the matrices A_{12} and A_{21}). Nonlinearity means here multi-linearity, since all equations are $m+n$-linear. Sun-Yuan Kung et al claim that this equation set always has solutions, which is consistent with the famous, but still unproved, Sontag conjecture (Sontag (1978)). Remarks on solving such equation sets are given in Section 3.4 here.

Note again that the concept of minimality for 2D/nD realizations is much more complex than in the 1D case. In particular, a state-space realization for a 2D/nD system can be controllable and observable but not minimal. Consider a proper, coprime, n-variate transfer function with a denominator whose degree in each variable s_i is equal to $t_i, i = 1,2,...,n$. Then by 1D theory, the minimal realization should have degree $N = \sum_{i=1}^{n} t_i$. However, such a realization does not necessarily exist even over the complex number field! In the 2D/nD case, the minimal existing realization can, in fact, have dimension greater than N and is termed a least order realization. A realization of dimension N is termed absolutely minimal.

Kung et al (1977) and Sontag (1978), who consider the fact that there is no rational procedure for constructing the minimal realization of a 2D linear system, both note that for some real bivariate rational functions, only a complex absolutely minimal realization can exist. An example is

$$\frac{s_1 + s_2}{s_1 s_2 - 1}.$$

Note here that application of the EOA procedure together with inversion of the variables produces the least order realization of dimension 3×3 but this is not absolutely minimal.

Both Kung et al (1977) and Sontag (1978) provide a solution for the very important case when the denominator polynomial is separable - a common occurrence in digital filtering applications. This case is also comparatively simple for MIMO 2D systems and for nD systems in general. Various construction techniques (based on various starting points and mathematical approaches) for these cases have been reported, e.g. Gu, Aravena and Zhou (1991), Raghuramireddy and Unbehauen (1993) and Hwang, Guo and Shieh (1993). Note also that the special case of separable denominators can (in principle at least) be extended to the general non-separable case by first applying appropriate feedback action to make the denominators separable, see, for example, Kawakami (1990).

This approach does not work in all cases because (as expected) there are some conditions, which must be satisfied for the necessary feedback action to exist.

To compare the Kung et al approach to the EOA, we use as a comparison the initial polynomial matrix representation $a_f(s_1, s_2)$. As shown in Chapter 4, a 1×1 matrix $\left[a_f(s_1, s_2)\right]$ guarantees that we can achieve a standard (nonsingular) realization for a principal polynomial, but only guarantees that we reach a subset of all possible equivalent state space realizations. (The 2×2 and higher cases require the solution of nonlinear algebraic equations). However, although the number of equations, which is equal to the number of transfer function coefficients, remains the same, the number of indeterminates can increase in comparison to the standard Kung *et al* method.

This last fact is a function of the initial representation used. For example, that used in (3.17) has

$$\varepsilon_{kl} = n + 1 + m + 1 + (n - k + 1)(l + 1) + (m - l + 1)(k + 1)$$

$$\tag{8.9}$$

$$= 2n + 2m + 4 + nl + mk - 2kl$$

coefficients / indeterminates and for $n=2$, $m=2$, $k=1$, $l=1$ $\varepsilon_{kl} = 10$ whilst the number of variables in the Kung *et al* approach is $2mn=8$. Also, the degree of the equation set nonlinearity decreases, which means that the EOA requires the solution of bilinear equations to construct the companion matrix of a bivariate or trilinear polynomial equations for a full state space realization. In the method of Kung *et al*, however, the equations which have to be solved are $n + m$ - linear and this can make the task easier.

It is important to note, however, that the problem of constructing the minimal realization of a 2D, and nD, linear system is still an open question - see Fornasini (1991). This is essentially due to the fact that the polynomial ring in more than one indeterminate is non-Euclidean (and is not even a principal ideal ring!). As a result, all unimodular matrices cannot be formed from products of elementary operations. This problem arises in the current context when augmenting the matrix size, *cf* Suslin (1976). For example, it is shown in and Park and Woodburn (1995) that although the so-called Cohn matrix,

$$F = \begin{bmatrix} 1 + xy & x^2 \\ -y^2 & 1 - xy \end{bmatrix},$$

$$\tag{8.10}$$

which is a special case of the so-called secondary operations (Sebek (1988)), cannot be expressed as the product of elementary operations, the augmented matrix

$$\tilde{F} = \begin{bmatrix} 1 & 0 & 0 \\ \hline 0 & 1 + xy & x^2 \\ 0 & -y^2 & 1 - xy \end{bmatrix}$$

$$\tag{8.11}$$

can. This fact was also noted by Gałkowski (1994). Moreover, it is valid for all unimodular 2D matrices, and the Woodburn, Park algorithm is an effective tool for finding a chain of corresponding elementary operations. Also, one can conjecture that this algorithm (appropriately modified if necessary) can be used in order to decrease the final dimension of 2D state-space realizations derived via the EOA.

The next essential contribution to the problem of the construction of state-space realizations for SISO and MIMO 2D systems was given by Kaczorek (1985). The main idea in the approaches described there is that the problem is considered over the one-variate polynomial ring with coefficients, which are one-variate polynomials in the second variable. Thus, instead of the ring $R[s_1, s_2]$ he considers the ring $R[s_1][s_2]$ or $R[s_2][s_1]$. In what follows, this method is revised.

In particular, return to the two-variate proper transfer function matrix obtained by setting $n=2$ in (2.34) which is clearly related to the Roesser state space model of (2.1)-(2.2) with

$$A = \begin{bmatrix} A_{11} & A_{12} \\ A_{21} & A_{22} \end{bmatrix}, \quad B = \begin{bmatrix} B_1 \\ B_2 \end{bmatrix}, \quad C = \begin{bmatrix} C_1 & C_2 \end{bmatrix}, \quad D, \tag{8.12}$$

where the block dimensions are with respect to $s_1 I_{t_1} \oplus s_2 I_{t_2}$. This transfer function matrix can be represented in the form of (6.20) with the least common denominator. The next step is to introduce the so-called first-level realizations, where all the entries are rational functions in one of the variables. In particular, note that the transfer function matrix $\aleph(s_1, s_2)$ can be written as

$$\aleph(s_1, s_2) = C(s_2) \left[s_1 I_{t_1} - A(s_2) \right]^{-1} B(s_2) + D(s_2), \tag{8.13}$$

where the matrices

$$A(s_2) = A_{11} + A_{12} \left[s_2 I_{t_2} - A_{22} \right]^{-1} A_{21},$$

$$B(s_2) = B_1 + A_{12} \left[s_2 I_{t_2} - A_{22} \right]^{-1} B_2,$$

$$C(s_2) = C_1 + C_2 \left[s_2 I_{t_2} - A_{22} \right]^{-1} A_{21},$$

$$D(s_2) = D + C_2 \left[s_2 I_{t_2} - A_{22} \right]^{-1} B_2 \tag{8.14}$$

define the first-level realisation. This result can also be found in Eising (1978, 1979 and 1980).

Alternatively, the proper rational matrix $\aleph(s_1, s_2)$ can be rewritten as

$$\aleph(s_1,s_2) = \frac{N(s_1,s_2)}{d(s_1,s_2)}$$

(8.15)

$$= \frac{N_{r_1}(s_2)}{d_{r_1}(s_2)} + \frac{\overline{N}_{r_1-1}(s_2)s_1^{r_1-1} + \cdots + \overline{N}_1(s_2)s_1 + \overline{N}_0(s_2)}{s_1^{r_1} + \overline{d}_{r_1-1}(s_2)s_1^{r_1-1} + \cdots + \overline{d}_1(s_2)s_1 + \overline{d}_0(s_2)},$$

where

$$N(s_1,s_2) = N_{r_1}(s_2)s_1^{r_1} + N_{r_1-1}(s_2)s_1^{r_1-1} + \cdots + N_1(s_2)s_1 + N_0(s_2)$$

$$d(s_1,s_2) = d_{r_1}(s_2)s_1^{r_1} + d_{r_1-1}(s_2)s_1^{r_1-1} + \cdots + d_1(s_2)s_1 + d_0(s_2)$$

$$N_j(s_2) \in \mathrm{Mat}_{q,p}(R[s_2]), \quad d_j(s_2) \in R[s_2], \quad j = 0,1,\ldots,r_1$$

$$\overline{d}_i(s_2) = \frac{d_i(s_2)}{d_{r_1}(s_2)} \in R(s_2),$$

$$\overline{N}_i(s_2) = \frac{N_i(s_2) - N_{r_1}(s_2)\overline{d}_i(s_2)}{d_{r_1}(s_2)} \in \mathrm{Mat}_{q,p}(R(s_2)), \quad i = 0,1,\ldots,r_1-1 \qquad (8.16)$$

The first-level realization can be derived by many well-known ways using standard algorithms for the 1D case, see the references in (Kaczorek (1985)). One choice is

$$A(s_2) = \begin{bmatrix} 0 & 0 & \cdots & 0 & -\overline{d}_0(s_2)I_q \\ I_q & 0 & \cdots & 0 & -\overline{d}_1(s_2)I_q \\ 0 & I_q & \cdots & 0 & -\overline{d}_2(s_2)I_q \\ \cdots & \cdots & \cdots & \cdots & \cdots \\ 0 & 0 & \cdots & 0 & -\overline{d}_{r_1-2}(s_2)I_q \\ 0 & 0 & \cdots & I_q & -\overline{d}_{r_1-1}(s_2)I_q \end{bmatrix}, \quad B(s_2) = \begin{bmatrix} \overline{N}_0(s_2) \\ \overline{N}_1(s_2) \\ \overline{N}_2(s_2) \\ \vdots \\ \overline{N}_{r_1-2}(s_2) \\ \overline{N}_{r_1-1}(s_2) \end{bmatrix},$$

$$C(s_2) = \begin{bmatrix} 0 & 0 & \cdots & 0 & I_q \end{bmatrix}, \quad D(s_2) = \left[\frac{N_{r_1}(s_2)}{d_{r_1}(s_2)}\right], \qquad (8.17)$$

and, it is straightforward to show that

$$\begin{bmatrix} A(s_2) & B(s_2) \\ C(s_2) & D(s_2) \end{bmatrix} = \begin{bmatrix} A_{12} \\ C_2 \end{bmatrix}\left[s_2 I_{t_2} - A_{22}\right]^{-1}\begin{bmatrix} A_{21} & B_2 \end{bmatrix} + \begin{bmatrix} A_{11} & B_1 \\ C_1 & D \end{bmatrix}. \qquad (8.18)$$

Now, note that

$$\lim_{s_2 \to \infty} \left[s_2 I_{t_2} - A_{22} \right]^{-1} = 0 , \tag{8.19}$$

which, in turn, yields

$$\lim_{s_2 \to \infty} \begin{bmatrix} A(s_2) & B(s_2) \\ C(s_2) & D(s_2) \end{bmatrix} = \begin{bmatrix} A_{11} & B_1 \\ C_1 & D \end{bmatrix} . \tag{8.20}$$

This equation when combined with (8.17) allows us to derive the submatrices $\{A_{11}, B_1, C_1, D\}$. Hence, applying one of the standard 1D realization procedures to the transfer function matrix

$$\overline{\aleph}(s_2) \triangleq \begin{bmatrix} A(s_2) & B(s_2) \\ C(s_2) & D(s_2) \end{bmatrix} - \begin{bmatrix} A_{11} & B_1 \\ C_1 & D \end{bmatrix} \tag{8.21}$$

yields the remaining state-space model submatrices, *i.e.* $\{A_{12}, A_{21}, A_{22}, B_2, C_2\}$. This is because

$$\overline{\aleph}(s_2) = \begin{bmatrix} A_{12} \\ C_2 \end{bmatrix} \left[s_2 I_{t_2} - A_{22} \right]^{-1} \begin{bmatrix} A_{21} & B_2 \end{bmatrix} . \tag{8.22}$$

Example 8.1 To illustrate the above procedure, Kaczorek gives the example

$$\aleph(s_1, s_2) = \frac{s_1 s_2 + 2}{s_1^2 s_2^2 + s_1^2 + s_1 s_2 + 1} ,$$

for which the 6×6 realization

$$A = \begin{bmatrix} 0 & 1 & 0 & 0 & 0 & 0 \\ 0 & 0 & -1 & 0 & 0 & -1 \\ 0 & 0 & 0 & 1 & 0 & 0 \\ 1 & 0 & -1 & 0 & 0 & 0 \\ 0 & 0 & 0 & 0 & 0 & 1 \\ 0 & 1 & 0 & 0 & -1 & 0 \end{bmatrix}, \quad B = \begin{bmatrix} 0 \\ 1 \\ 0 \\ 0 \\ 0 \\ 0 \end{bmatrix}, \quad C = \begin{bmatrix} 0 & 0 & 2 & 0 & 0 & 1 \end{bmatrix}, \quad D = \begin{bmatrix} 0 \end{bmatrix}$$

has been derived.

It is easy to check that using the EOA approach in this case yields a lower dimensional result, *i.e.* apply

```
> with(linalg):
> Fsp:=matrix(1,1,[z*(s^2*p^2+s^2+s*p+1)-s*p-2]);
> B1:=Augment(Fsp);
> B1:=addcol(B1,1,2,-z);
> B1:=map(simplify,addrow(B1,1,2,s^2*p^2+s^2+s*p+1));
```

```
> B2:=Augment(B1);
> B2:=addrow(B2,1,3,s);
> B2:=map(simplify,addcol(B2,1,2,-s*p^2-s-p));
> B2:=map(simplify,addcol(B2,1,3,p));
> B3:=Augment(B2);
> B3:=addrow(B3,1,2,p);
> B3:=map(simplify,addcol(B3,1,3,s*p+1));
> B3:=addcol(B3,1,4,-1);
> B4:=Augment(B3);
> B4:=addcol(B4,1,4,s);
> B4:=addrow(B4,1,2,-p);
> B4:=addrow(B4,1,3,1);
> B4:=mulrow(B4,2,-1);
> B4:=mulrow(B4,4,-1);
> pr:=[1,5,2,3,4]:      #permutation for rows
> pc:=[4,3,1,2,5]: #permutation for columns,
```

which yields the following 4-dimensional realization

$$
A = \begin{bmatrix} 0 & 0 & -1 & 0 \\ -1 & 0 & 0 & 0 \\ \hline 1 & 0 & 0 & 1 \\ 0 & -1 & -1 & 0 \end{bmatrix}, \quad B = \begin{bmatrix} 0 \\ 2 \\ \hline -1 \\ 0 \end{bmatrix}, \quad C = \begin{bmatrix} 1 & 0 & 0 & 0 \end{bmatrix}, \quad D = \begin{bmatrix} 0 \end{bmatrix},
$$

which is clearly absolutely minimal. Kaczorek (1985) also gives various canonical realizations for use in either the MIMO or SISO cases.

The next very significant contribution to the state space realization problem for both MIMO and SISO 2D linear systems was given by Żak et al (1986). This is also based on replacing the two variable polynomial ring by a one variable polynomial/rational function ring in one of the variables with coefficients which are rational functions/polynomials in the second variable. The main result of Żak et al is that they introduce into 2D systems theory the concepts of modal controllability and observability, see also Klamka (1988), and associate an absolutely minimal realization to these concepts. Note, however, that modal controllability and observability are stronger concepts in 2D as opposed to 1D linear systems. Also so-called local controllability and observability of a state space realization does not guarantee its minimality.

The intrinsic features of the Żak et al approach are that they

a. determine a minimal first order realization,

    b.   find the minimum dimension of the second-level realization by using the notion of the Mac Millan degree, and

    c.   use elementary operations over a field of rational functions to find the second-level, final realization.

To givemore details, let $\{A(s_2), B(s_2), C(s_2), D(s_2)\}$ be the first-level minimal realization of dimension χ for a transfer function matrix $\aleph(s_1, s_2)$, which means that the entries of the matrix are represented in the form of $\aleph(s_1)(s_2)$ *i.e.* they are treated as rational functions in s_2 with coefficients which are rational functions in s_1. Next, they propose finding minimal dimension δ and construct the matrix

$$\begin{bmatrix} I_\delta & 0 & 0 \\ 0 & s_1 I_\chi - A(s_2) & -B(s_2) \\ 0 & -C(s_2) & -D(s_2) \end{bmatrix} \tag{8.23}$$

to yield the required minimal second-level realization by using elementary operations.

This method is very interesting but there is one critical point to note. This is that the elementary operations are elementary over $R(s_1)$ - the ring of rational functions in s_1. In such a case, it is not easy to determine the required operations and there are no algorithms for constructing the second level realization from the first. As a result, only simple examples can be treated. This problem is avoided in the EOA approach by associating the ring of bivariate rational functions with the ring of trivariate polynomials. Also it can be conjectured that all methods based on the representing the transfer function matrix $\aleph(s_1, s_2)$ in the form $\aleph(s_1)(s_2)$ or $\aleph(s_2)(s_1)$ will be rather difficult to generalize relative to the EOA approach in the cases when $n > 2$.

Interesting results based on the same idea, but without the use of rational elementary operations, and characterized by matrix operations performed in one stage (not successively as in the EOA) are given by Pugh, McInerney, Boudellioua, Hayton (1998) and by Kaczorek (1999). In these cases, however, singular realizations are the final result.

The last method we refer to in more detail, is the developed by Cockburn and Morton (1997). This work provides an interesting new application in nD systems, see also Doyle (1982), Beck, Doyle, Glover (1996) for characterization of system uncertainty. Let $\Delta = \operatorname{diag}\left(q_1 I_{r_1}, \ldots, q_p I_{r_p}\right)$ represent the system uncertainty. Then the input-output mapping of the feedback system with Δ in the loop can be expressed in the form

$$S(\Delta, M) \triangleq M_{22} + M_{21}\Delta(I - M_{11}\Delta)^{-1} M_{12}, \tag{8.24}$$

which is equivalent to (7.3). Hence, the following important problem arises:

Given a p-variate polynomial matrix, which represents the input-output mapping $S(\Delta, M)$, find a state-space realization

$$M = \begin{bmatrix} M_{11} & M_{12} \\ M_{21} & M_{22} \end{bmatrix}, \tag{8.25}$$

possibly of the minimal order.

Cockburn and Morton (1997) attempt to apply the standard circuit approach, where systems are assumed to be connections of simpler systems. Hence, the system is decomposed into sums and products of simple uncertain subsystems, for which a minimal realization can be readily obtained. Finally, these subsystems are combined together to obtain a representation of the original system. The authors develop a specific tensor product based theory of possible decompositions and implement it in the form of a MATLAB® based numerical tool. As shown in Section 3.1 of this monograph, the example given by Cockburn and Morton can be solved by the EOA with the result of a smaller dimension for the final realization.

The state-space realizations for 2D/nD linear systems have been the subject of much work based on a large range of approaches. It would be an impossible task to review all of this and instead we give an overview of a representative range. Note also that much of the research in this area can be found in the circuits/signal processing literature since the problem of realization/synthesis of nD filters is of considerable practical (and theoretical) significance.

In essence, all of this work can be sub-divided into

a. methods, which limit attention to important special cases with the aim of finding the minimal or `near minimal' solution, and

b. methods for the general case.

The first grouping here includes methods for systems with separable denominators and circuits based methods. These include the following:

a. orthogonal MIMO 2D filters (Piekarski (1988)),

b. allpass filters (Gisladottir, Lev-Ari and Kailath (1990), Varoufakis, Antoniou (1991))

c. 2D multi-ports (MIMO systems) described by a scattering matrix (Kummert (1989))

d. SISO systems, which can be expanded into multi-variate continued fractions (Mentzelopoulous, Theodorou (1991))

e. lattice filters (Antoniou (1993))

It is also worth of mentioning methods related to finite memory and related systems (Fornasini, Zampieri (1991)).

The second grouping contains the following methods:

a. all the methods discussed previously in this overview (except the separable-denominator case)

b. an interesting method given by Tzafestas, Theodorou (1987) , where an attempt is made to develop an inductive approach to the construction of a state-space realization for SISO n-D systems

c. the subject of this monograph – the EOA

d. the direct method by Gałkowski (1991), which is briefly summarized below

e. the use of three-way matrices (Gałkowski (1981b)).

Note now that the methods previously discussed or mentioned refer to the Roesser model. There are also methods that attempt to produce the Fornasini-Marchesini models, which can also be treated as an introductory step to obtaining the Roesser model (Kung (1977), Section 3.5 of this monograph), see Hinamoto, Fairman (1984) for the so-called Attasi model.

It is also necessary to note that there are approaches that aim to adapt classic 1D techniques to the state-space description of 2D systems. The first such attempt was made by Porter and Aravena (1984) and, independently, by Kaczorek (1987). An intrinsic feature of these approaches is that the so-called global state, input, output vectors over the separation set are defined as, for example,

$$X(n) \triangleq \mathrm{col}(y(n,0), y(n-1,1), \ldots, y(0,n)). \tag{8.26}$$

The resulting models are characterized by a variable structure since it is easy to see that the dimension of global vectors increases as 'n' increases.

In the remainder of this chapter we consider methods for constructing state space realizations for 2D/nD linear systems under the following headings:

a. construction of an *a priori* non-minimal realization for nD SISO systems,

b. the use of multi-way matrices in the realization of n-variate polynomial matrices

c. a direct method for construction of minimal nD realizations.

8.2 *A priori* Non-minimal Realization

As already noted here, the result of Kung *et al* generalizes easily to an arbitrary nD transfer function, *i.e.* to an arbitrary SISO nD system, (Gałkowski (1981a), (1991)). Here, we briefly summarize the general case. Start by writing the polynomial a_f as follows:

$$a_f(s_1, \cdots, s_{n+1}) = \sum_{j_1=0}^{t_1} \sum_{j_2=0}^{t_2} \cdots \sum_{j_{n+1}=0}^{t_{n+1}(=1)} a_{j_{n+1}j_n\cdots j_1} \prod_k^{n+1} s_k^{t_k-j_k}. \tag{8.27}$$

Now, let

$$b_{j_{n+1}j_n\cdots j_2 0} \triangleq a_{j_{n+1}j_n\cdots j_2 0}, \tag{8.28}$$

and

$$b_{j_{n+1}j_n\cdots j_2 1} \triangleq a_{j_{n+1}j_n\cdots j_2 1} - a_{j_{n+1}j_n\cdots j_2 0} a_{00\cdots 0_1}, \tag{8.29}$$

where

$$j_k = 0,1,\cdots,t_k, \iota_k = 1,2,\cdots,t_k.$$

Next, construct the following blocks:

$$B_1 = \begin{bmatrix} 1 & 0 & \cdots & 0 \end{bmatrix}^T,$$

$$B_2 = \begin{bmatrix} \mathbf{b}_{0\cdots\alpha_2 j_1} & \mathbf{b}_{0\cdots\alpha_3 \iota_2 j_1} & \cdots & \mathbf{b}_{j_{n+1}0\cdots\alpha_2 j_1} \end{bmatrix},$$

$$B_3 = \begin{bmatrix} \mathbf{b}_{0\cdots\iota_3 0 j_1} & \mathbf{b}_{0\cdots\alpha_4 \iota_3 j_2 j_1} & \cdots & \mathbf{b}_{\iota_{n+1}0\cdots\alpha_3 j_2 j_1} \end{bmatrix},$$

$$\cdots\cdots\cdots\cdots\cdots$$

$$B_k = \begin{bmatrix} \mathbf{b}_{0\cdots\iota_k 0\cdots 0 j_1} & \mathbf{b}_{0\cdots\alpha_{k+1}\iota_k j_{k-1}\cdots j_1} & \cdots & \mathbf{b}_{\iota_{n+1}0\cdots\alpha_k j_{k-1}\cdots j_1} \end{bmatrix},$$

$$\cdots\cdots\cdots\cdots\cdots$$

$$B_n = \begin{bmatrix} \mathbf{b}_{\alpha_n 0\cdots 0 j_1} & \mathbf{b}_{\iota_{n+1}\iota_n j_{n-1}\cdots j_1} \end{bmatrix},$$

$$B_{n+1} = \begin{bmatrix} \mathbf{b}_{\iota_{n+1}0\cdots 0 j_1} \end{bmatrix}, \tag{8.30}$$

where

$$\mathbf{b}_{\alpha j_1} = \begin{bmatrix} b_{\alpha 0} & b_{\alpha 1} & \cdots & b_{\alpha t_1} \end{bmatrix}^T. \tag{8.31}$$

Now, partition the matrix B_k as

$$B_k = \begin{bmatrix} B_{k1} \\ B_{k2} \end{bmatrix}, \tag{8.32}$$

where B_{k1} is the first row of B_k, and hence, construct the $N \times N$ matrix

$$H = \begin{bmatrix}
B_{12} & B_{22} & B_{32} & \cdots & B_{n+1,2} \\
H_2^1 & -B_{21}^2 & -B_{31}^2 & \cdots & -B_{n+1,1}^2 \\
 & H_2^2 & & & \\
H_3^1 & -B_{21}^3 & -B_{31}^3 & \cdots & -B_{n+1,1}^3 \\
 & H_3^2 & & & \\
\cdots & \cdots & \cdots & \cdots & \cdots \\
H_n^1 & -B_{21}^n & -B_{31}^n & \cdots & -B_{n+1,1}^n \\
 & H_n^2 & & & \\
H_{n+1}^1 & -B_{21}^{n+1} & -B_{31}^{n+1} & \cdots & -B_{n+1,1}^{n+1}
\end{bmatrix}, \tag{8.33}$$

where

$$N = \sum_{l=1}^{n+1} \mu_l \,.$$

(8.34)

The blocks used here are defined as

$$B_{12} = \begin{bmatrix} -a_{0\cdots01} & 1 & 0 & \cdots & 0 \\ -a_{0\cdots02} & 0 & 1 & \cdots & 0 \\ \vdots & & & \ddots & \\ -a_{0\cdots0t_1-1} & 0 & 0 & \cdots & 1 \\ -a_{0\cdots0t_1} & 0 & 0 & \cdots & 0 \end{bmatrix}, \quad H_k^1 = \begin{bmatrix} -1 & 0 \\ \hline 0 & 0 \end{bmatrix}, \quad B_{k1}^k = \begin{bmatrix} B_{k1} \\ \hline -I_{t_k-1} & 0 \end{bmatrix},$$

$$B_{k1}^m = \begin{bmatrix} B_{k1} \\ 0 \end{bmatrix}, \qquad k, m = 2,3,\ldots, n+1; k \neq m.$$

(8.35)

Note here that the matrix H_k^1 contains only one nonzero element, which is placed in the left upper corner. It still remains to determine the blocks H_k^2, $k=2,\ldots,n$, which are of the dimension $(\mu_k - t_k) \times N$. Assume first that blocks B_{k2} are of dimension $t_1 \times \mu_k$, where

$$\mu_k = t_k + \sum_{j=k+1}^{n+1} t_k t_j \prod_{l=2}^{k-1} (t_l + 1)$$

(8.36)

and

$$\prod_{l=2}^{k-1} (t_l + 1) = 1 \text{ for } k = 2$$

(8.37)

and $\mu_{n+1} = t_{n+1} = 1$. In what follows, $H_k^2, k = 2,3,\ldots,n$ is the $(\mu_k - t_k) \times N$ zero-one matrix defined as follows:

Consider the v-th row of $H_k^2, k = 2,3,\ldots,n$, $\nu = 1,2,\ldots,\mu_k - t_k$. This row is set equal to the w-th row of H. Let the wth column of H contains as a sub-column, the following column of $B \triangleq \begin{bmatrix} B_{12} & B_{22} & \cdots & B_{n+1,2} \end{bmatrix}$:

$$\begin{bmatrix} b_{\alpha 1} & b_{\alpha 2} & \cdots & b_{\alpha t_1} \end{bmatrix}^T,$$

(8.38)

where α denotes the word index

$$\alpha \triangleq j_{n+1} j_n \cdots j_{k+1} j_k j_{k-1} \cdots j_2 \,.$$

(8.39)

Finally, element 1 of the considered row must lie at the intersection with the column of H that contains the sub-column

$$\begin{bmatrix} b_{\beta 1} & b_{\beta 2} & \cdots & b_{\beta t_1} \end{bmatrix}^T,$$ (8.40)

where

$$\beta \triangleq j_{n+1} j_n \cdots j_{k+1} j_k - 1 j_{k-1} \cdots j_2.$$ (8.41)

All the remaining elements of the considered row are zero.

Note now that the matrix H is the companion matrix of the augmented polynomial

$$a'_f(s_1, \cdots, s_{n+1}) = \prod_{l=2}^{n} s_l^{P_l} a_f(s_1, \cdots, s_{n+1}),$$ (8.42)

i.e. related to the transfer function

$$f'(s_1, \cdots, s_{n+1}) = \frac{\prod_{l=2}^{n} s_l^{P_l} a(s_1, \cdots, s_{n+1})}{\prod_{l=2}^{n} s_l^{P_l} b(s_1, \cdots, s_{n+1})},$$ (8.43)

where $p_k = \mu_k - t_k$, *i.e.*

$$a'_f(.) = \det\left[\bigoplus_{k=1}^{n+1} s_k I_{\mu_k} - H \right].$$ (8.44)

Note also that the final dimension of H can be rewritten as

$$N = \sum_{l=1}^{n+1} \mu_l = 1 + t_1 + \sum_{i=2}^{n} 2t_i + \sum_{i=2}^{n-1} \sum_{j=i+1}^{n} 2t_i t_j$$

$$+ \sum_{i=2}^{n-2} \sum_{j=i+1}^{n-1} \sum_{k=j+1}^{n} 2t_i t_j t_k + \cdots + 2t_2 t_3 \cdots t_n.$$ (8.45)

It is straightforward to see that considerable redundancy exists here, *i.e.*

$$R = \sum_{i=2}^{n} t_i + \sum_{i=2}^{n-1} \sum_{j=i+1}^{n} 2t_i t_j + \sum_{i=2}^{n-2} \sum_{j=i+1}^{n-1} \sum_{k=j+1}^{n} 2t_i t_j t_k + \cdots + 2t_2 t_3 \cdots t_n,$$ (8.46)

which rapidly increases when the number of variables increases.

Example 8.2 Consider the case when $n = 4$, $t_1 = t_3 = t_3 = 2$; $t_4 = 1$. The matrix H here can now be written as

$$H = \begin{bmatrix} B_{12} & B_{22} & B_{32} & B_{42} \\ H_2^1 & -B_{21}^2 & -B_{31}^2 & -B_{41}^2 \\ & H_2^2 & & \\ H_3^1 & -B_{21}^3 & -B_{31}^3 & -B_{41}^3 \\ & H_3^2 & & \\ H_4^1 & -B_{21}^4 & -B_{31}^4 & -B_{41}^4 \end{bmatrix},$$

(8.47)

where

$$B_{12} = \begin{bmatrix} -a_{0001} & 1 \\ -a_{0002} & 0 \end{bmatrix}, \quad H_2^1 = \begin{bmatrix} -1 & 0 \\ 0 & 0 \end{bmatrix},$$

$$H_2^2 = \begin{bmatrix} 0 & 0 & 0 & 0 & 0 & 0 & 0 & 0 & 0 & 0 & 1 & 0 & 0 & 0 & 0 & 0 & 0 & 0 & 0 & 0 \\ 0 & 0 & 0 & 0 & 1 & 0 & 0 & 0 & 0 & 0 & 0 & 0 & 0 & 0 & 0 & 0 & 0 & 0 & 0 & 0 \\ 0 & 0 & 0 & 0 & 0 & 0 & 0 & 0 & 0 & 0 & 0 & 1 & 0 & 0 & 0 & 0 & 0 & 0 & 0 & 0 \\ 0 & 0 & 0 & 0 & 0 & 0 & 1 & 0 & 0 & 0 & 0 & 0 & 0 & 0 & 0 & 0 & 0 & 0 & 0 & 0 \\ 0 & 0 & 0 & 0 & 0 & 0 & 0 & 0 & 0 & 0 & 0 & 0 & 0 & 0 & 0 & 0 & 0 & 0 & 0 & 1 \\ 0 & 0 & 0 & 0 & 0 & 0 & 0 & 0 & 1 & 0 & 0 & 0 & 0 & 0 & 0 & 0 & 0 & 0 & 0 & 0 \end{bmatrix},$$

$$B_{22} = \begin{bmatrix} b_{0011} & b_{0021} & b_{0111} & b_{0121} & b_{0211} & b_{0221} & b_{1011} & b_{1021} \\ b_{0012} & b_{0022} & b_{0112} & b_{0122} & b_{0212} & b_{0222} & b_{1012} & b_{1022} \end{bmatrix},$$

$$H_3^1 = \begin{bmatrix} -1 & 0 \\ 0 & 0 \end{bmatrix}, \quad H_4^1 = \begin{bmatrix} -1 & 0 \end{bmatrix},$$

$$-B_{21}^2 = \begin{bmatrix} -a_{0010} & -a_{0020} & -a_{0110} & -a_{0120} & -a_{0210} & -a_{0220} & -a_{1010} & -a_{1020} \\ 1 & 0 & 0 & 0 & 0 & 0 & 0 & 0 \end{bmatrix},$$

$$-B_{21}^3 = \begin{bmatrix} -a_{0010} & -a_{0020} & -a_{0110} & -a_{0120} & -a_{0210} & -a_{0220} & -a_{1010} & -a_{1020} \\ 0 & 0 & 0 & 0 & 0 & 0 & 0 & 0 \end{bmatrix},$$

$$-B_{21}^4 = \begin{bmatrix} -a_{0010} & -a_{0020} & -a_{0110} & -a_{0120} & -a_{0210} & -a_{0220} & -a_{1010} & -a_{1020} \end{bmatrix},$$

$$B_{32} = \begin{bmatrix} b_{0101} & b_{0201} & b_{1101} & b_{1111} & b_{1121} & b_{1201} & b_{1211} & b_{1221} \\ b_{0102} & b_{0202} & b_{1102} & b_{1112} & b_{1122} & b_{1202} & b_{1212} & b_{1222} \end{bmatrix},$$

$$-B_{31}^2 = \begin{bmatrix} -a_{0100} & -a_{0200} & -a_{1100} & -a_{1110} & -a_{1120} & -a_{1200} & -a_{1210} & -a_{1220} \\ 0 & 0 & 0 & 0 & 0 & 0 & 0 & 0 \end{bmatrix},$$

$$-B_{31}^3 = \begin{bmatrix} -a_{0100} & -a_{0200} & -a_{1100} & -a_{1110} & -a_{1120} & -a_{1200} & -a_{1210} & -a_{1220} \\ 1 & 0 & 0 & 0 & 0 & 0 & 0 & 0 \end{bmatrix},$$

$$-B_{31}^4 = \begin{bmatrix} -a_{0100} & -a_{0200} & -a_{1100} & -a_{1110} & -a_{1120} & -a_{1200} & -a_{1210} & -a_{1220} \end{bmatrix},$$

$$H_3^2 = \begin{bmatrix} 0 & 0 & 0 & 0 & 0 & 0 & 0 & 0 & 0 & 0 & 0 & 0 & 0 & 0 & 0 & 0 & 0 & 1 \\ 0 & 0 & 0 & 0 & 0 & 0 & 0 & 0 & 1 & 0 & 0 & 0 & 0 & 0 & 0 & 0 & 0 & 0 \\ 0 & 0 & 0 & 0 & 0 & 0 & 0 & 0 & 0 & 1 & 0 & 0 & 0 & 0 & 0 & 0 & 0 & 0 \\ 0 & 0 & 0 & 0 & 0 & 0 & 0 & 0 & 0 & 0 & 0 & 0 & 1 & 0 & 0 & 0 & 0 & 0 \\ 0 & 0 & 0 & 0 & 0 & 0 & 0 & 0 & 0 & 0 & 0 & 0 & 0 & 1 & 0 & 0 & 0 & 0 \\ 0 & 0 & 0 & 0 & 0 & 0 & 0 & 0 & 0 & 0 & 0 & 0 & 0 & 0 & 1 & 0 & 0 & 0 \end{bmatrix},$$

$$B_{42} = \begin{bmatrix} b_{1001} \\ b_{1002} \end{bmatrix}, \quad B_{41}^2 = B_{41}^3 = \begin{bmatrix} a_{1000} \\ 0 \end{bmatrix}, \quad B_{41}^4 = \begin{bmatrix} -a_{1000} \end{bmatrix}. \tag{8.48}$$

This companion matrix can be transformed to the canonical Smith form by using elementary operations, which is not generally the case for nD systems. This canonical Smith form can be written as

$$\aleph_\hbar = \left[\begin{array}{c|c} I_\varepsilon & 0 \\ \hline 0 & \begin{array}{c} \Psi \\ \hline \chi(s_1, s_2, \cdots, s_n) \mid a_f(s_1, s_2, \cdots, s_{n+1}) \end{array} \end{array} \right], \tag{8.49}$$

where

$$\varepsilon = \sum_{i=1}^{n+1} \mu_i - \sum_{k=2}^{n} \sum_{\substack{m=2 \\ m \geq k}}^{n} \xi_{km} - 1, \tag{8.50}$$

$$\xi_{km} = \begin{cases} t_{m+1} & k = 2 \\ \\ t_{k-1} t_{m+1} & k = 3 \\ \\ t_{k-1} t_{m+1} \prod_{l=2}^{k-2} (t_l + 1) & k = 4, 5, \cdots, n \end{cases}, \tag{8.51}$$

$$\Psi = \mathrm{diag}\Big\{ s_2{}^{t_2} I_{\xi_{22}}, s_3{}^{t_3} I_{\xi_{33}}, \cdots, s_n{}^{t_n} I_{\xi_{nn}}, s_2{}^{t_2} s_3{}^{t_3} I_{\xi_{23}}, s_3{}^{t_3} s_4{}^{t_4} I_{\xi_{34}}, \cdots$$

$$, \qquad (8.52)$$

$$s_k{}^{t_k} s_{k+1}{}^{t_{k+1}} \cdots s_n{}^{t_n} I_{\xi_{kn}}, \cdots, s_2{}^{t_2} s_3{}^{t_3} \cdots s_n{}^{t_n} I_{\xi_{kn}} \Big\}$$

and $\chi(s_1, s_2, \cdots, s_n)$ is some polynomial row which depends on the particular choice of elementary operations applied.

It should be noted that in the group of extensions to the method of Kung et al (1977), the method developed by Gałkowski is characterized by a relatively simple structure, which is particularly attractive to the application of computer algebra tools. It also has the significant advantage that its Smith canonical form can be derived by elementary operations (recall that for multivariate polynomial matrices there may exist unimodular matrices, which are not the product of elementary operations).

8.3 Multi-way Companion Matrices

One of the numerous methods proposed to obtain easier solutions to the 2D/nD realization problem was the use of so-called multi-way matrices. This is related to the fact the nonlinear equations, which determine a system matrix in this are usually over-determined, i.e. there are more equations than variables. In particular, a flat (multi-way terminology) square matrix has too few entries in comparison to the number of coefficients in a multivariate transfer function or polynomial. The means of improving this situation is based on enlarging the number, which is already the case when we first extend the state space dimension, and hence the use of non-minimal realizations, and secondly when we use n-way matrices (here cubes).

Here we next give a short description of the basic ideas of multi-way matrices. This is followed by the construction of a three-dimensional (3-way) companion matrix for a three-variate polynomial is developed. The idea then is to use this as a basis for obtaining a minimal realization of an arbitrary two-variate transfer function - see Gałkowski (1981b).

8.3.1 The History of Multi-way Matrices

Multi-way matrices are, essentially, generalization of certain flat ones. The entries of commonly used flat matrices are displayed in a linear 2-dimensional space Z^2, whereas the entries of multi-way matrices are displayed in a linear p-dimensional space Z^p. Such matrices have attracted attention of mathematicians since the end of XIX century. According to the Baraniecki (1879) university textbook on determinant theory, the first people who introduced three-way, cube matrices and their determinants were Vandermonde [Paris 1871] and Somov [St. Petersbourg 1864]. Oldenburger (1940), Rice and Scott have also obtained the essential extensions. For detailed references see, for example, Zieliński (1990).

There is now a well-established algebra for multi-way matrices. This includes methods to solve multi-way matrix equations, calculation of their determinants, eigenvalues, and inverses (when defined). For a detailed treatment of multi-way matrix algebra see, for example, Sokolov (1972).

As regards applications, there have been some attempts to exploit this theory. For example, Kron (1965) applied multi-way matrix methods to the analysis of circuits, Zieliński *et al* (1990) applied cube permanents to state identification of a power system. Multi-way matrices also play a significant role in multivariate polynomial theory and possibly in applications to nD systems theory. For example, a 3-way companion matrix for an arbitrary trivariate principal polynomial was formulated and proved using the theory of three-way determinants by Gałkowski 1981b). Such a representation of a trivariate polynomial is very helpful in the analysis of its properties such as separability and factorizability. Next, multi-way matrices were employed to solve the problem of the separability of the general multi-variate polynomial, Gałkowski (1992).

Next we develop a new multi-way matrix description of multi-variate polynomials and show how it permits easy solutions of a number of problems such as linear transformations of polynomial variables (Gałkowski (1993)), the Caley-Hamilton theorem for nD linear systems, and the factorization of multivariate polynomials (Gałkowski (1996c)).

8.3.2 Multi-way Matrices - Mathematical Basics

Let P be some field of numbers. Any array of $t_1 \times t_2 \times \cdots \times t_p$ elements $A_{i_1 i_2 \cdots i_p}$ $(i_j = 1,2,\ldots,t_j, j = 1,2,\ldots,p)$ of a field P, obtained by placing $A_{i_1 i_2 \cdots i_p}$ at the point $(i_1,i_2,\ldots,i_p)$ in a linear p-dimensional space, defines a p-way matrix A over P. Hence, we can introduce a mutually exclusive, exhaustive partition $T_1, T_2, \ldots, T_r$ of the set of the indices of A. Then the matrix A can be represented as the display $(A_{T_1 T_2 \cdots T_r})$, which is the r-way matrix $(r \leq p)$ with the indices $T_1, T_2, \ldots, T_r$. It is easy to see that on setting $r = 2$, we can write the p-way matrix as an ordinary (2-way) matrix. For example, if $A = \left[A_{i_1 i_2 i_3}\right], i_1, i_2, i_3 = 1,2$ and $T_1 = \{i_3\}, T_2 = \{i_1,i_2\}$, then

$$A = \left[A_{T_1 T_2}\right] = \begin{bmatrix} A_{111} & A_{121} & A_{211} & A_{221} \\ A_{112} & A_{122} & A_{212} & A_{222} \end{bmatrix},$$ (8.53)

where the vertical line separates the i_1-layers of A i.e.

$$\left[A_{\bar{i}_1 i_2 i_3}\right]; \bar{i}_1 \in \{1,2\}; i_2, i_3 = 1,2.$$ (8.54)

Standard operations such as summation and multiplication by a scalar can obviously be defined. Hence, let

$$\mathbf{B} = \left[B_{T_1 T_2}\right] = \begin{bmatrix} B_{111} & B_{121} & B_{211} & B_{221} \\ B_{112} & B_{122} & B_{212} & B_{222} \end{bmatrix}.$$

Then

$$\mathbf{C} = \mathbf{A} + \mathbf{B} = \left[C_{T_1 T_2}\right]$$

$$= \begin{bmatrix} A_{111} + B_{111} & A_{121} + B_{121} & A_{211} + B_{211} & A_{221} + B_{221} \\ A_{112} + B_{112} & A_{122} + B_{122} & A_{212} + B_{212} & A_{222} + B_{222} \end{bmatrix} \tag{8.55}$$

and

$$\mathbf{D} = \lambda \mathbf{A} = \left[\lambda A_{T_1 T_2}\right] = \begin{bmatrix} \lambda A_{111} & \lambda A_{121} & \lambda A_{211} & \lambda A_{221} \\ \lambda A_{112} & \lambda A_{122} & \lambda A_{212} & \lambda A_{222} \end{bmatrix}. \tag{8.56}$$

It is also possible to define the determinant of such matrices, which is a crucial point for possible applications. First, for the p-way matrix A define the so-called transversal

$$A_{i_1^1 i_2^1 \dots i_p^1}, A_{i_1^2 i_2^2 \dots i_p^2}, \dots, A_{i_1^t i_2^t \dots i_p^t}, \tag{8.57}$$

where the values $\left\{i_j^1, i_j^2, \dots, i_j^t\right\}$ of the index $i_j (j = 1, 2, \dots, p)$ form some permutation of the set $\{1, 2, \dots, t\}$. Next, introduce the following expression:

$$K = (-1)^{\sum_{k=1}^m I_k} A_{i_1^1 i_2^1 \dots i_p^1} A_{i_1^2 i_2^2 \dots i_p^2} \cdots A_{i_1^t i_2^t \dots i_p^t}, \tag{8.58}$$

where I_k is the number of the inversions in the permutation $\left\{i_k^1, i_k^2, \dots, i_k^t\right\}$, $(k = 1, 2, \dots, m = 2\alpha \leq p)$. These k indices of $\mathbf{A}$, where k is always even, are called "alternating" or "signant" and the remaining indices are called "nonalternating" or "nonsignant". Now, following Oldenburger (1940), and Sokolov (1972), see also Gałkowski (1981b), the determinant of A can be defined as

$$\det \mathbf{A}_{\substack{+/-+/- \ +/- \ + \ + \\ i_1 \ i_2 \ \dots \ i_m \ i_{m+1} \dots i_p}} = \sum K \tag{8.59}$$

where +/- signifies alternation, and + signifies non-alternation, and the summation is over the whole set of the transversals of $\mathbf{A}$. For example, the determinant of the matrix A given by (1) can be expressed as follows:

$$\det \mathbf{A} = A_{111} A_{222} - A_{112} A_{211} + A_{122} A_{211} - A_{121} A_{212}, \tag{8.60a}$$
$$i_1 - \text{nonalternating}$$

$$\det \mathbf{A} = A_{111} A_{222} - A_{112} A_{211} - A_{122} A_{211} + A_{121} A_{212}, \tag{8.60b}$$
$$i_2 - \text{nonalternating}$$

$$\det A = A_{111}A_{222} + A_{112}A_{211} - A_{122}A_{211} - A_{121}A_{212} ,$$
$$i_3 - \text{nonalternating}$$

(8.60c)

and

$$\det A = A_{111}A_{222} + A_{112}A_{211} + A_{122}A_{211} + A_{121}A_{212} .$$
all indices are nonalternating

(8.60d)

This last determinant form is termed "pemanent".

8.3.3 Three-way Companion Matrix of a Trivariate Polynomial

The natural definition of a companion matrix for a trivariate polynomial given as

$$a(s,p,z) = \sum_{h=0}^{r}\sum_{k=0}^{m}\sum_{l=0}^{n} a_{hkl} s^{n-l} p^{m-k} z^{r-h}$$

(8.61)

is

$$\det\left[sI_n \oplus pI_m \oplus zI_r - H\right] = a(s,p,z),$$

(8.62)

where $\oplus$ denotes the direct sum. As has been noted before, the flat matrix H has got too few entries (variables) for easy solution of (8.62). It is possible, however, to avoid this difficulty by developing the three-way companion matrix. Accordingly, suppose that the companion matrix for the polynomial given in (8.61) is defined as:

$$\det\left[sI_n^3 \oplus pI_m^3 \oplus zI_r^3 - G^3\right] = a(s,p,z)$$

(8.63)

where $I_k^3 (k = m, n, r)$ are the $k \times k \times k$ three-way cube "unit" matrices

$$I_k^3 = \begin{bmatrix} 1 & 0 & \cdots & 0 & 0 & 0 & \cdots & 0 & & 0 & 0 & \cdots & 0 \\ 0 & 0 & \cdots & 0 & 0 & 1 & \cdots & 0 & & 0 & 0 & \cdots & 0 \\ \cdots & \cdots & \cdots & \cdots & \cdots & \cdots & \cdots & \cdots & \cdots & \cdots & \cdots & \cdots & \cdots \\ 0 & 0 & \cdots & 0 & 0 & 0 & \cdots & 0 & & 0 & 0 & \cdots & 1 \end{bmatrix},$$

(8.64)

and G is the $(m+n+r) \times (m+n+r) \times (m+n+r)$ three-way matrix. Thus the equality of polynomials (8.62) is equivalent to a system of $(m+1)(n+1)(r+1)-1$ equations in $(m+n+r)^3$ variables. Now, it is clear that the number of variables is always greater than the number of equations, which makes the construction simpler. As an example, we show the cube companion matrix G^3 for the polynomial with $r = m = n = 2$ that can be defined by the following $\{\{i_3\}, \{i_1, i_2\}\}$ – display :

$$G^3 = \left[g_{i_1 i_2 i_3}\right] = \begin{bmatrix} -a_{001} & 0 & 0 & 0 & -\hat{a}_{101} & -\hat{a}_{201} & 0 & 1 & 0 & 0 & 0 & 0 \\ -a_{002} & 0 & 0 & 0 & -\hat{a}_{102} & -\hat{a}_{202} & 0 & 0 & 0 & 0 & 0 & 0 \\ 1 & 0 & 0 & 0 & 0 & 0 & 0 & 0 & 0 & 0 & 0 & 0 \\ 0 & 0 & 0 & 0 & 0 & 0 & 0 & 0 & 0 & 0 & 0 & 0 \\ 0 & 0 & 1 & 1 & 0 & 0 & 0 & 0 & 0 & 0 & 0 & 0 \\ 0 & 0 & 0 & 0 & 0 & 0 & 0 & 0 & 0 & 0 & 0 & 0 \end{bmatrix}$$

$$\begin{vmatrix} 0 & 0 & -\hat{a}_{011} & 0 & -\tilde{a}_{111} & -\tilde{a}_{211} & 0 & 0 & 0 & -\hat{a}_{021} & -\tilde{a}_{121} & -\tilde{a}_{221} \\ 0 & 0 & -\hat{a}_{012} & 0 & -\tilde{a}_{112} & -\tilde{a}_{212} & 0 & 0 & 0 & -\hat{a}_{022} & -\tilde{a}_{122} & -\tilde{a}_{222} \\ 0 & 0 & -a_{010} & 0 & 0 & 0 & 0 & 0 & 0 & -a_{020} & 0 & 0 \\ 0 & 0 & 1 & 0 & 0 & 0 & 0 & 0 & 0 & 0 & 0 & 0 \\ 0 & 0 & 0 & 0 & -1 & -1 & 0 & 0 & 0 & 0 & -1 & -1 \\ 0 & 0 & 0 & 0 & 0 & 0 & 0 & 0 & 0 & 0 & 0 & 0 \end{vmatrix}$$

$$\begin{vmatrix} 0 & 0 & 0 & 0 & 0 & 0 & 0 & 0 & 0 & 0 & 0 & 0 \\ 0 & 0 & 0 & 0 & 0 & 0 & 0 & 0 & 0 & 0 & 0 & 0 \\ 1 & 0 & -\hat{a}_{110} & -\hat{a}_{120} & 0 & 0 & 1 & 0 & -\hat{a}_{210} & -\hat{a}_{220} & 0 & 0 \\ 0 & 0 & 0 & 0 & 0 & 0 & 0 & 0 & 0 & 0 & 0 & 0 \\ 1 & 0 & 0 & 0 & -a_{100} & 0 & 1 & 0 & 0 & 0 & 0 & -a_{200} \\ 0 & 0 & 0 & 0 & 1 & 0 & 0 & 0 & 0 & 0 & 0 & 0 \end{vmatrix} \tag{8.65}$$

where the vertical lines delimit the $\{i_1\}$-layers of G^3 and

$$\hat{a}_{ij0} \triangleq a_{ij0} - a_{i00}a_{0j0}, \hat{a}_{i0k} \triangleq a_{i0k} - a_{i00}a_{00k}, \hat{a}_{0jk} \triangleq a_{0jk} - a_{0j0}a_{00k},$$

$$\tilde{a}_{ijk} \triangleq a_{ijk} - a_{i00}a_{0jk} - a_{0j0}a_{i0k} - a_{00k}a_{ij0} + 2a_{i00}a_{0j0}a_{00k}. \tag{8.66}$$

For details see Gałkowski (1981b). It should be stressed that such a cube companion matrix is suitable for analysing some properties of trivariate polynomials such as separability.

The above discussion has shown that a three-way companion matrix exists for each trivariate polynomial. Unfortunately, however, this result cannot be directly applied to obtain a state space realization of a two-variate transfer function due, essentially, to the fact that the three-way matrix G^3 is not an automorphism on the vector space. Such matrices can be applied to, for example, bilinear or square systems but not linear. Despite this difficulty, the three-way companion matrix is useful for the characterization of three-variate polynomials. For example, Gałkowski (1981b) has used them to give necessary and sufficient conditions for polynomial separability.

8.4 Minimal Realization for Multi-linear n-variate Transfer Functions

In this section, the direct solution of the absolute minimal state-space realization for n-D systems is again discussed. As already noted, solving the polynomial equation (2.33), where $p = q = 1$ in (2.34), is very difficult and also, there are difficulties in deriving appropriate, numerically feasible tests for the existence of solutions. The problem of solving multivariate polynomial equations has been discussed at numerous points in this monograph, e.g. Section 3.4 and Chapter 4. It turns out, however, that for the particular case of SISO nD systems, where both the denominator and numerator of the transfer function are linear polynomials in each of their variables i.e. $t_1 = t_2 = \cdots = t_n = 1$, one can derive conditions for the existence of a solution. In turn, this enables us to solve the underlying polynomial equations which is presented next in concise form.

Start by rewriting the polynomial a_f associated with the given transfer function in the form

$$a_f(s_1,\cdots,s_{n+1}) = \sum_{j_1=0}^{1} \sum_{j_2=0}^{1} \cdots \sum_{j_{n+1}=0}^{1} a_{j_{n+1}j_n\cdots j_1} \prod_k^{n+1} s_k^{1-j_k} . \tag{8.67}$$

For simplicity, the coefficients of a polynomial are denoted as

$$a_{\underbrace{\underbrace{0\cdots 0\,1\,0\,\cdots\,\cdots\,\cdots\,0\,1\,0\,\cdots\,0\,\overbrace{1\,0\,\cdots\,0}^{i_1}}_{i_2}}_{i_k}} := a_{i_1 i_2\cdots i_k} . \tag{8.68}$$

Now, the problem can be stated as follows: Given a multi-linear polynomial a_f find the matrix $H = [h_{ij}]_{n+1,n+1}$ that satisfies the polynomial equation (8.67), where, as stated before, $t_1 = t_2 = \cdots = t_n = 1$.

This polynomial equation can be rewritten in the form of an algebraic equation set with the entries h_{ij} of the matrix H as the unknowns i.e.

$$(-1)^k M_{i_1 i_2\cdots i_k} = a_{i_1 i_2\cdots i_k} , \tag{8.69}$$

$k \in \{1,2,\ldots,n+1\}$, $\{i_1,i_2,\ldots,i_k\} \in C_k\{1,2,\ldots,n+1\}$, where $M_{i_1 i_2\cdots i_k}$ denotes the main minor of the matrix H that consists of elements belonging to the intersections of $i_1 -$, $i_2 -,\ldots$, $i_k - th$ rows and columns of H and $C_k\{1,2,\ldots,n+1\}$ denotes the set of all the k-element combinations from the $n+1$-element set $\{1,2,\ldots,n+1\}$. To avoid misunderstanding, $\mathbf{M}_{i_1 i_2\cdots i_k}$ denotes here the respective sub-matrix and $M_{i_1 i_2\cdots i_k}$ - the value of the minor, i.e. $M_{i_1 i_2\cdots i_k} := \det \mathbf{M}_{i_1 i_2\cdots i_k}$.

To write the equation set (8.69) in a clearer form, define $\forall \{i_1, i_2, \ldots, i_k\} \in C_k \{1, 2, \ldots, n+1\}$ the set $\Re \{i_1, i_2, \ldots, i_k\}$ of all the permutations $r = \{j_1, j_2, \ldots, j_k\}$ of the set $\{i_1, i_2, \ldots, i_k\}$ such that

a. $j_1 = i_1$,

b. $j_2 \neq i_k$,

c. $j_k > j_2$.   (8.70)

Now, note that the value of $M_{i_1 i_2 \cdots i_k}$ consists of the sums of products of matrix H elements that lie at transversals of $M_{i_1 i_2 \cdots i_k}$, and some of them can be the direct sums of transversales belonging to mutually exclusive, exhaustive main sub-minors of $M_{i_1 i_2 \cdots i_k}$. Using (8.70) enables us to represent some transversales as functions of polynomial coefficients, which in turn, limits the number of the addends in (8.69). Hence, (8.69) can be rewritten in the form

$$\sum_{r \in \Re(i_1, i_2, \ldots, i_k)} \left(A_r + A_{r*} \right) = -\tilde{a}_{i_1, i_2, \ldots, i_k}$$ (8.71)

where $k \in \{1, 2, \ldots, n+1\}$, $\{i_1, i_2, \ldots, i_k\} \in C_k \{1, 2, \ldots, n+1\}$. Here only the transversales, which cannot be represented as the direct sums discussed above occur on the left-hand side of the equations. Now, introduce

$$A_r := h_{i_1 j_2} h_{j_2 j_3} \cdots h_{j_k i_1},$$

$$A_{r*} := h_{i_1 j_k} h_{j_k j_{k-1}} \cdots h_{j_2 i_1}$$ (8.72)

$\{j_1, j_2, \ldots, j_k\} \in \Re \{i_1, i_2, \ldots, i_k\}$ and

$$\tilde{a}_{i_1, i_2, \ldots, i_k} := \sum_{z \in \aleph(i_1, i_2, \ldots, i_k)} (-1)^{u-1} (u-1)! \prod_{v=1}^{u} a_{z_v}$$ (8.73)

where $\aleph \{i_1, i_2, \ldots, i_k\}$ is a set of all the mutually exclusive, exhaustive partitions of the set $\{i_1, i_2, \ldots, i_k\}$ and

$$z = \{z_1, \ldots, z_v, \ldots, z_w, \ldots, z_u\} \mid u = 1, 2, \ldots, k; \ z_v, z_w \subseteq \{i_1, i_2, \ldots, i_k\};$$

$$z_v \cap z_w = \emptyset; \bigcup_{v=1}^{u} z_v = \{i_1, i_2, \ldots, i_k\}.$$ (8.74)

Hence, if k_v denotes the number of the elements /strength/ of the set z_v then

$$\sum_{v=1}^{u} k_v = k.$$

Now, recall the form of equations (8.71) for k=1,2 and 3 $i.e.$:

k=1: $h_{ii} = -\tilde{a}_i := -a_i$, $i = 1,2,...,n+1$, (8.75)

k=2: $h_{i_1 i_2} h_{i_2 i_1} = -\tilde{a}_{i_1 i_2} := -a_{i_1 i_2} + a_{i_1} a_{i_2}$, (8.76)

k=3:

$$h_{i_1 i_2} h_{i_2 i_3} h_{i_3 i_1} + h_{i_1 i_3} h_{i_3 i_2} h_{i_2 i_1} = -\tilde{a}_{i_1 i_2 i_3} := -a_{i_1 i_2 i_3} + a_{i_1} a_{i_2 i_3}$$

(8.77)

$$+ a_{i_2} a_{i_1 i_3} + a_{i_3} a_{i_1 i_2} - 2 a_{i_1} a_{i_2} a_{i_3} ,$$

where $\{i_1, i_2\} \in C_2\{1,2,...,n+1\}$, $\{i_1, i_2, i_3\} \in C_3\{1,2,...,n+1\}$. Also, note that the set of equations of (8.71) is over-determined, $i.e.$ there are more equations than variables. This implies that the number of elements h_{ij}, i,j=1,2,...,n+1, of the matrix H is not sufficient to assure the satisfaction of (8.71) for all the possible coefficients $a_{i_1 i_2 \cdots i_k}$. Thus, not all multi-linear, multi-variate polynomials possess an absolutely minimal companion matrix.

Our aim here is to choose the set of basic equations whose solution will be given later. In what follows, we substitute them into the remaining equations in such a way as to eliminate variables. This, in effect, gives necessary and sufficient conditions for solvability of the equation set and hence for the existence of an absolutely minimal companion matrix. This method fails, however, if the polynomial is not multi-linear since it is then not easy to eliminate variables to explicitly give limitations. The reason is because a multi-linear polynomial equation can be solved uniquely with respect to each variable. Consider, for example, the equation

$$xyztv + 2xtzv - 3ztv - xv + 4tv - y + 1 = 0 .$$

Clearly, each variable can be uniquely obtained as a rational function of the remaining variables, e.g.

$$x = \frac{3ztv - 4tv + y - 1}{yztv + 2tzv - v - 1} .$$

For general polynomial equations this is impossible since such a substitution can be multi-valued or even may not exist. It can be conjectured that a very useful approach to such problems would be again to rely on using algebraic tools related to so-called Grobner bases (Buchberger (1970)). This is left here as a topic for future research.

Return now to solving the multi-linear problem and note that the equation (8.75) alone provides the result for diagonal elements of H, $i.e.$ h_{ii}. In what follows, the basic equations are derived from the set (8.76)-(8.77). To detail this, note that (8.77), based on (8.76), can be solved as

$$A_{i_1 i_2 i_3} := h_{i_1 i_2} h_{i_2 i_3} h_{i_3 i_1} = \frac{1}{2}\left[-\tilde{a}_{i_1 i_2 i_3} \pm \sqrt{\Delta_{i_1 i_2 i_3}} \right],$$ (8.78)

and

$$A_{i_1 i_2 i_3 *} := h_{i_1 i_3} h_{i_3 i_2} h_{i_2 i_1} = \frac{1}{2}\left[-\tilde{a}_{i_1 i_2 i_3} \mp \sqrt{\Delta_{i_1 i_2 i_3}} \right],$$ (8.79)

$$\Delta_{i_1 i_2 i_3} := \left(\tilde{a}_{i_1 i_2 i_3} \right)^2 + 4\tilde{a}_{i_1 i_2} \tilde{a}_{i_1 i_3} \tilde{a}_{i_2 i_3},$$ (8.80)

$\{i_1, i_2, i_3\} \in C_3 \{1, 2, \ldots, n+1\}$. Thus the problem can be reformulated as that of solving (8.76), (8.79) and (8.80). This equation set is still over-determined, *i.e.* there exist $A_{i_1 i_2 i_3}$ and $A_{i_1 i_2 i_3 *}$, which depend on the others. This leads in effect to some dependencies among the polynomial coefficients that must hold for a minimal companion matrix to exist.

In what follows, assume that $\forall \; \{i_1, i_2\} \in C_2 \{1, 2, \ldots, n+1\}$, $\tilde{a}_{i_1 i_2} \neq 0$. Then the following lemmas can be established.

Lemma 8.1 *The value of $A_{i_1 i_2 i_3 *}$ can be determined from $A_{i_1 i_2 i_3}$ as*

$$A_{i_1 i_2 i_3 *} = -\frac{\tilde{a}_{i_1 i_2} \tilde{a}_{i_1 i_3} \tilde{a}_{i_2 i_3}}{A_{i_1 i_2 i_3}}.$$ (8.81)

Lemma 8.2 $\forall \; \{i_1, i_2, i_3\} \in C_3 \{1, 2, \ldots, n+1\}$ *the quantity $A_{i_1 i_2 i_3}$ can be presented as*

$$A_{i_1 i_2 i_3} = \frac{A_{1 i_1 i_2} A_{1 i_2 i_3} \tilde{a}_{i_1 i_3}}{A_{1 i_1 i_3} \tilde{a}_{1 i_2}}.$$ (8.82)

Lemma 8.3 *Let* $\{j_1, j_2, \ldots, j_k\} \in \Re\{i_1, i_2, \ldots, i_k\}$, *where* $k \in \{1, 2, \ldots, n+1\}$, $\{i_1, i_2, \ldots, i_k\} \in C_k \{1, 2, \ldots, n+1\}$. *Then*

$$A_{j_1 j_2 j_3 \cdots j_k} = (-1)^{k-1} \frac{A_{j_1 j_2 j_3} A_{j_1 j_3 j_4} \cdots A_{j_1 j_{k-1} j_k}}{\tilde{a}_{j_1 j_3} \tilde{a}_{j_1 j_4} \cdots \tilde{a}_{j_1 j_{k-1}}}$$ (8.83)

Proofs. The proofs in each are immediate by inspection on using (8.75)-(8.79). ∎
Lemmas 8.1÷8.3 allow the following problem solution:

Theorem 8.1 *Assume* a_i; $\tilde{a}_{i_1i_2} \neq 0$ *and* $\tilde{a}_{1i_1i_2}$, $\{i_1,i_2\} \in C_2\{1,2,...,n+1\}$, *are arbitrary. Then, solve the respective equations (8.76)-(8.77), and using (8.82)-(8.83), construct the required conditions to be satisfied for the remaining polynomial coefficients.*

Proof. The proof is immediate on previous analysis. ∎

The following, next theorem can now be established.

Theorem 8.2 *The solution of the basic subset of (8.75)-(8.77), i.e. (8.75), (8.76) and the subset of (8.77) for* $\{i_1,i_2,i_3\} := \{1,i_1,i_2\}$, *can be accomplished as*

$$h_{i_1i_1} = h^*_{i_1i_1} = -a_{i_1}, \tag{8.84}$$

$$h_{1i_1} = h^*_{1i_1} \text{ (arbitrary value)}, \tag{8.85}$$

$$h_{i_11} = h^*_{i_11} = -\frac{\tilde{a}_{1i_1}}{h^*_{1i_1}}, \tag{8.86}$$

$$h_{i_1i_2} = h^*_{i_1i_2} = -\frac{A_{1i_1i_2}h^*_{1i_2}}{\tilde{a}_{1i_2}h^*_{1i_1}}, \tag{8.87}$$

$$h_{i_2i_1} = h^*_{i_2i_1} = -\frac{\tilde{a}_{i_1i_2}}{h_{i_1i_2}}, \tag{8.88}$$

$$i_1,i_2 = 1,2,...,n+1, \ i_1 < i_2.$$

Proof. An immediate consequence of (8.75)-(8.80).

The only problem that now remains is that of consistency within (8.87) and (8.88). In particular, note that, as for $h_{i_1i_2}$, which can be derived from (8.87), the value of $h_{i_2i_1}$ can be obtained as

$$h_{i_2i_1} = h^*_{i_2i_1} = -\frac{A_{1i_1i_2}h^*_{1i_1}}{\tilde{a}_{1i_1}h^*_{1i_2}}, \tag{8.89}$$

c.f. (8.79). The question now arises: Are (8.88) and (8.89) consistent? It is easy to see that, due to (8.78)-(8.79) and (8.81), we have

$$h^*_{i_1i_2}h^*_{i_2i_1} = \frac{A_{1i_1i_2}A_{1i_1i_2}\cdot h^*_{1i_1}h^*_{1i_2}}{\tilde{a}_{1i_1}\tilde{a}_{1i_2}h^*_{1i_1}h^*_{1i_2}} = -\tilde{a}_{i_1i_2},$$

i.e. (8.76) holds. Note here that $A_{i_1 i_2 i_3}$ are the known values given by (8.78)-(8.79). ∎

The problem of completing the construction of the necessary and sufficient existence conditions is now discussed. First consider the equations of (8.77) for $i_1 \neq 1$. Due to (8.81)-(8.82), we have that

$$\tilde{a}_{i_1 i_2 i_3} = \frac{A_{1 i_1 i_2} A_{1 i_2 i_3} \tilde{a}_{i_1 i_3}}{A_{1 i_1 i_3} \tilde{a}_{1 i_2}} - \frac{\tilde{a}_{i_1 i_2} \tilde{a}_{1 i_2} \tilde{a}_{i_2 i_3} A_{1 i_1 i_3}}{A_{1 i_1 i_2} A_{1 i_2 i_3}}. \tag{8.90}$$

Note also that possible values of $A_{1 i_1 i_2}$, $A_{1 i_1 i_3}$, $A_{1 i_2 i_3}$ belong to the two-element sets

$$\left\{ \frac{1}{2}\left[-\tilde{a}_{1 i_1 i_2} + \sqrt{\Delta_{1 i_1 i_2}}\right], \frac{1}{2}\left[-\tilde{a}_{1 i_1 i_2} - \sqrt{\Delta_{1 i_1 i_2}}\right] \right\},$$

$$\left\{ \frac{1}{2}\left[-\tilde{a}_{1 i_1 i_3} + \sqrt{\Delta_{1 i_1 i_3}}\right], \frac{1}{2}\left[-\tilde{a}_{1 i_1 i_3} - \sqrt{\Delta_{1 i_1 i_3}}\right] \right\},$$

or

$$\left\{ \frac{1}{2}\left[-\tilde{a}_{1 i_2 i_3} + \sqrt{\Delta_{1 i_2 i_3}}\right], \frac{1}{2}\left[-\tilde{a}_{1 i_2 i_3} - \sqrt{\Delta_{1 i_2 i_3}}\right] \right\}, \tag{8.91}$$

respectively. Thus, $\tilde{a}_{i_1 i_2 i_3}$ can have values from a set of $8(=2^3)$ possible elements obtained from (8.80) by applying all the possible combinations of (8.91). In a similar fashion, the set of the remaining conditions for $3 < k < n+1$ can be derived from Lemma 8.3 with regard (8.91). Thus, for given basic coefficients a_i, $\tilde{a}_{i_1 i_2} \neq 0$ and $\tilde{a}_{1 i_1 i_2}$, $\{i_1, i_2\} \in C_2\{1,2,...,n+1\}$ there exist a finite number $(2^\alpha$, where $\alpha = \binom{n}{2})$ of acceptable sets for the remaining coefficients values. Also, note that the choice of one of the two possibilities for $A_{1 i_1 i_2}$, *c.f.* (8.91), also influences the resulting matrix H. Such a change results in the interchange of the values between (8.87) and (8.88), and hence in partial transposition in the matrix H. To illustrate the above procedure consider the following example.

Example 8.3 Consider a 4-variate polynomial

$$a_f(s_1,s_2,s_3,s_4) = s_1 s_2 s_3 s_4 + a_1 s_2 s_3 s_4 + a_2 s_1 s_3 s_4 + a_3 s_1 s_2 s_4 + a_4 s_1 s_2 s_3$$

$$+ a_{12} s_3 s_4 + a_{13} s_2 s_4 + a_{14} s_2 s_3 + a_{23} s_1 s_4 + a_{24} s_1 s_3 \tag{8.92}$$

$$+ a_{34} s_1 s_2 + a_{123} s_4 + a_{124} s_3 + a_{134} s_2 + a_{234} s_1 + a_{1234}$$

The previous analysis allows us to assume that $a_1 = a^*_1$, $a_2 = a^*_2$, $a_3 = a^*_3$, $a_{12} = a^*_{12}$, $a_{13} = a^*_{13}$, $a_{14} = a^*_{14}$, $a_{23} = a^*_{23}$, $a_{24} = a^*_{24}$, $a_{34} = a^*_{34}$, $a_{123} = a^*_{123}$, $a_{124} = a^*_{124}$, $a_{134} = a^*_{134}$ are arbitrary, where $*$ denotes the given value of a variable. Theorem 8.2 yields that

$$h_{11} = -a_1, \quad h_{22} = -a_2, \quad h_{33} = -a_3, \quad h_{44} = -a_4,$$

$$h_{12} = h^*_{12}, \quad h_{13} = h^*_{13}, \quad h_{14} = h^*_{14} \text{ (arbitrary values)},$$

$$h_{21} = -\frac{\tilde{a}_{12}}{h^*_{12}}, \quad h_{31} = -\frac{\tilde{a}_{13}}{h^*_{13}}, \quad h_{41} = -\frac{\tilde{a}_{14}}{h^*_{14}},$$

$$h_{23} = h^*_{23} = -\frac{A_{123} h^*_{13}}{\tilde{a}_{13} h^*_{12}}, \quad h_{24} = h^*_{24} = -\frac{A_{124} h^*_{14}}{\tilde{a}_{14} h^*_{12}},$$

$$h_{34} = h^*_{34} = -\frac{A_{134} h^*_{14}}{\tilde{a}_{14} h^*_{13}}, \quad h_{32} = -\frac{\tilde{a}_{23}}{h^*_{23}},$$

$$h_{42} = -\frac{\tilde{a}_{24}}{h^*_{24}}, \quad h_{43} = -\frac{\tilde{a}_{34}}{h^*_{34}} \tag{8.93}$$

where the values of $A_{123}, A_{124}, A_{134}$ belong to the two-element sets given by (8.78)-(8.79), i.e.

$$\left\{ \frac{1}{2}\left[-\tilde{a}_{123} + \sqrt{\Delta_{123}} \right], \frac{1}{2}\left[-\tilde{a}_{123} - \sqrt{\Delta_{123}} \right] \right\},$$

$$\left\{ \frac{1}{2}\left[-\tilde{a}_{124} + \sqrt{\Delta_{124}} \right], \frac{1}{2}\left[-\tilde{a}_{124} - \sqrt{\Delta_{124}} \right] \right\},$$

or

$$\left\{ \frac{1}{2}\left[-\tilde{a}_{134} + \sqrt{\Delta_{134}} \right], \frac{1}{2}\left[-\tilde{a}_{134} - \sqrt{\Delta_{134}} \right] \right\}. \tag{8.94}$$

Hence, there are 8 acceptable sets for the coefficients a_{234}, a_{1234}, and hence, a_{234} has to satisfy

$$\tilde{a}_{234} = \frac{A_{123} A_{134} \tilde{a}_{14}}{A_{124} \tilde{a}_{13}} - \frac{\tilde{a}_{23} \tilde{a}_{13} \tilde{a}_{34} A_{124}}{A_{123} A_{134}}. \tag{8.95}$$

The value of a_{1234} has to satisfy the following condition:

$$\tilde{a}_{1234} = h_{12}h_{23}h_{34}h_{41} + h_{14}h_{43}h_{32}h_{21} + h_{12}h_{24}h_{43}h_{31}$$

$$+ h_{13}h_{34}h_{42}h_{21} + h_{13}h_{32}h_{24}h_{41} + h_{14}h_{42}h_{23}h_{31}$$

which can be rewritten as, (use (8.83)),

$$\tilde{a}_{1234} = -\frac{A_{123}A_{134}}{a_{13}} - \frac{A_{123^*}A_{134^*}}{a_{13}} - \frac{A_{124}A_{134^*}}{a_{13}}$$

$$\hspace{3cm} (8.96)$$

$$- \frac{A_{124^*}A_{134}}{a_{13}} - \frac{A_{123^*}A_{124}}{a_{13}} - \frac{A_{123}A_{124^*}}{a_{13}},$$

where

$$\bar{a}_{1234} \triangleq a_{1234} - a_1 a_{234} - a_2 a_{134} - a_3 a_{124} - a_4 a_{123} - a_{12}a_{34} - a_{13}a_{24}$$

$$- a_{14}a_{23} + 2a_1 a_2 a_{34} + 2a_1 a_3 a_{24} + 2a_1 a_4 a_{23} + 2a_2 a_3 a_{14} \hspace{1cm} (8.97)$$

$$+ 2a_2 a_4 a_{13} + 2a_3 a_4 a_{12} - 6a_1 a_2 a_3 a_4.$$

Up to now we have assumed that all the coefficients $\tilde{a}_{i_1 i_2}$ are nonzero. If this is not the case the conditions developed previously are generally only sufficient. To expand on this problem, introduce first the following notation:

$$\Gamma := \left\{ \{i_1, i_2\} \in C_2 \{1, 2, \ldots, n+1\} : \tilde{a}_{i_1 i_2} = 0 \right\}. \hspace{1cm} (8.98)$$

It is easy to note that if Γ is not empty and $\{i_1, i_2\} \in \Gamma$, then only one of the variables $h_{i_1 i_2}$ or $h_{i_2 i_1}$ has to be equal to zero in order for the corresponding equation of (21) to hold. Now, define the second set

$$\Delta := \{ \{i_1, i_2\} \in C_2 \{1, 2, \ldots, n+1\} : \forall k = 1, 2, \ldots, n+1, k \neq i_1, k \neq i_2;$$

$$\hspace{3cm} (8.99)$$

$$\tilde{a}_{i_1 k} = 0 \vee (\wedge) \, \tilde{a}_{k i_2} = 0; \, \exists k_1 : \tilde{a}_{i_1 k_1} = 0 \Leftrightarrow \exists k_2 : \tilde{a}_{k_2 i_2} = 0 \}$$

The set Δ plays a crucial role in the further analysis, since if Δ is empty, then the way of proceeding developed in the previous section for the case of the empty set Γ is still valid and the existence conditions defined previously are still necessary and sufficient. If the set Δ is not empty, then the limitations on the empty set Γ can be relaxed in the sense that the former necessary and sufficient conditions become only sufficient. Crucially, it can be seen from the definition of Δ that the set Γ must have at least n-1 elements for the set Δ to be non-empty. Also, note that when the set Δ is non-empty, the analysis becomes very complicated. Hence we seek means to transform the problem characterised by a non-empty set Δ to an equivalent problem characterised by an empty set Δ. Such a possibility relies on using the well-known multi-variate, generalised, bilinear transform, or the variable inversion (Gałkowski (1992)). Thorough analysis of this problem can be found in Gałkowski (2001).

9. Conclusions and Further Work

The problem of how to construct state space realizations for a given 2D/nD linear system written, for example, in 2D/nD transfer function (or transfer function matrix) form is central to various applications of such systems in control and many other areas. As such, it has received considerable attention in the relevant literature. In this monograph, substantial results have been developed on this fundamental problem which, unlike the 1D linear systems case, is still an open question. Results on the equivalence properties of such realizations and their relationships to causality and minimality have also been developed.

A key feature of 2D/nD linear systems is the number of distinct state space systems, which can be used to describe their dynamics. The most commonly encountered state space descriptions are those due to Roesser and Fornasini-Marchesini respectively. It has been shown here that from an algebraic standpoint, described using polynomial matrices, that the Fornasini-Marchesini model can be interpreted as an intermediate result (or step) in the construction of a Roesser model for the underlying dynamics. Also, it is a simple exercise to extend these state space models to other forms, which contain the main characteristics of both of them.

This monograph has also detailed an approach to the realization problem for 2D/nD linear systems based, in the SISO case, on a multivariate polynomial setting. The essence of this method is to replace the given nD transfer function is replaced by a polynomial in $n + 1$ variables. Also it has been shown how to generalize this approach to the MIMO and singular systems cases.

The construction of state space realizations for 2D/nD linear systems is very important in terms of applications since it is then possible (at least in principle) to `optimize' the structure used. For example, this can lead to structures characterized by a minimal number of specified elements. In the 1D case, it is only necessary to invoke the concept of similarity and hence attention can be restricted to a subset of all possible transformations in order to preserve specified key systems properties across different realizations. For example, if orthogonal transformations are used then these preserve loselessness, passivity, and symmetry in circuit realizations.

In the 2D/nD case, it is not so easy to use similarity transformations and for this key reason, Gałkowski and others (see the relevant cited references in this monograph) have developed key results on the use of the so-called Elementary Operations Algorithm (EOA) in this area. The development of this approach has formed the starting point (after the necessary background material has been given) for this monograph. In effect, this algorithm produces a range of state-space realizations, in the nD case, by a well defined sequence of operations applied to an equivalent (to the defining transfer function (matrix)) polynomial (matrix) in $n + 1$ indeterminates together with augmentation of its dimensions.

Often when modelling physical systems, it is impossible to construct standard equations. Instead, only a subset the so-called singular (or descriptor or implicit) equations can be obtained. An immediate implication of this fact is that non-monic polynomials can appear in transfer function (matrix) descriptions. This is a well-known fact in `classical' (1D) circuit theory but its importance increases markedly in the 2D/nD case due, for example, to its clear links to causality. Causality is a key feature of `classical' (1D) systems since it must be present for physical realizability and in this case the general concept of `time' imposes a natural ordering into `past', `present' and future and this leads to physical realizability.
Again, the situation is different in the 2D/nD case where, in effect, some of the indeterminates have a spatial, rather than temporal, characteristic. Hence causality (as a generalization of the 1D case) is not required since it is only necessary to be able to recursively perform the required computations.

An immediate consequence of this last fact re causality is that singularity is much more important in the 2D/nD case. Note, however, that singularity may not be an intrinsic feature of the system but can, for example, be introduced by `non-adequate' identification of the model or by discretization methods employed
as part of, for example, simulation studies. Also this feature can be avoided in some cases by application of well-defined algebraic operations.

Singularity has also formed a significant topic in this monograph, where it has been demonstrated that this undesirable feature can be avoided in some cases by reformulating the state vector or changing the model type. Also it has been shown that there are direct links between different standard and singular models, which further confirms the fluent boundary between causality and non-causality in nD linear systems. Some new types of singularity/causality have also been developed.

It is important to note that application of EOA does not always produce standard (or nonsingular) realizations - even for the class of principal transfer functions. In particular, some additional relationships between the numerators and denominators of such transfer functions involved and these can be difficult to identify these during the various stages of applying the EOA. Also, it has been established that a necessary condition for the existence of a nonsingular system matrix of the Roesser form of a transfer function matrix is that all entries have principal numerator polynomials - otherwise only a singular solution to this problem exists.

The fact that it is possible to achieve essentially singular realizations starting from principal transfer functions (or matrices) is a considerable limitation of the method. It has been shown here, however, that it is possible to improve the EOA method in such a way as to obtain standard (nonsingular) solutions for a much wider class of principal transfer functions (or matrices) than with its basic version. This has been achieved by appropriate use of variable transformations such as inversion and the generalized bilinear transformation (which have a long history in systems theory).

A critical performance comparator of the various algorithms developed is the minimality of the resulting system matrix. In which context, it is important to note that EOA does not always yield a minimal realization and, in fact, a positive answer here strongly depends on the order in which the individual operations are performed. For example, in the SISO case, the final result of a minimal realization depends on the initial polynomial matrix representation of the given nD transfer function. Here it has been shown that there exist representations, which guarantee the existence of minimal representations and research is proceeding to obtain conditions, which guarantee the existence of such a representation (mainly for the general nD case).

In MIMO examples, the realizations obtained at the end of the first stage of the EOA are often singular, but this difficulty can be removed by variable inversion of use of the bilinear transformation. Note, however, that the final realizations obtained are mostly redundant, *i.e.* non-minimal. A key point, however, is that very often the level of redundancy introduced by EOA is less than that of other methods (see Section 8.1). Hence only EOA operation chains which produce state space realizations, which provide the minimum possible dimension should be used.

Chapter 8 here (at the end) has given a critical overview of the EOA in comparison to other methods. This has made use of early work in this area by the author (and others) and, in particular, the following.

a. Methods for direct construction of *a priori* non-minimal realizations of SISO nD systems.

b. The development of methods for constructing minimal realizations of n-variate polynomial matrices using results from the theory of multi-way matrices.

c. Direct methods for constructing a minimal nD realization for a first-degree multivariate polynomial.

In general, the work reported in this work has confirmed and/or established the following main advantages of the EOA approach.

a. Its relative simplicity and flexibility of application.

b. The possibility of obtaining a much wider class of equivalent realizations compared to other methods.

c. The possibility of obtaining relatively low dimensional realizations compared to other methods.

d. The possibility of making effective use of symbolic computations (here Maple has been exclusively used).

There are, however, a number of areas in which profitable research can still be undertaken. These include those discussed next.

First there is the possibility of using coprime polynomial matrix representations. Consider first the general 2D coprime representation $F(s_1, s_2) = N(s_1, s_2)D(s_1, s_2)^{-1}$, constructed using, for example, the method of Guiver and Bose (1982). Then, it can be conjectured that realizations with greatly reduced dimensionality would be obtained. This is an extremely promising direction for further research in the subject area of this monograph for both the 2D and nD case where it should be noted that the latter case is, in general, much more complicated (in relative terms).

Other promising areas for further research include the following.

a. The use of Grobner bases (Buchberger (1970)) to solve the problem of finding appropriate initial representations.

b. The use of Suslin stability theory (Suslin (1976)) based equivalence algorithms (Park, Woodburn (1995)).

c. The use of the general EOA approach (modified as necessary) over the so-called multivariate Laurent polynomials.

d. The extension to the nonlinear case.

e. The application of the EOA in the study of linear repetitive processes and other applications areas.

References

Amann N., Owens D.H., Rogers E. (1998) Predictive optimal iterative learning control. *Int. J. Control*, Vol. 69, No. 2, 203-226

Ansell H.G. (1964) On certain two-variable generalizations of circuits theory, with applications to network of transmission lines and lumped reactances, *IEEE Trans. on Circuits and Systems*, Vol. CAS-11, 214-223

Antoniou G.E. (1993) 2-D discrete time lossless bounded real functions: Minimal state space realizations, *Electronics Letters*, Vol. 29, No.23, 2008-2009

Aplevich J.D. (1974) Direct computation of canonical forms for linear systems by elementary matrix operations, *IEEE Trans. on Automatic Control*, April, 124-126.

Baraniecki M. A., (1879) Teorya Wyznaczników (Determinantów) (Theory of Determinants), Paris 1879, The Kórnik Library, in Polish

Badreddin E., Mansour M. (1983) A multivariable normal form for model reduction of discrete time systems, *Systems Control Lett.*, Vol. 2, 271-285; (1984) A second multivariable normal form for model reduction of discrete time systems, *Systems Control Lett.*, Vol. 4, 109-117

Beck C.L., Doyle J., Glover K. (1996) Model reduction of multidimensional and uncertain systems, *IEEE Trans. on Automatic Control*, Vol.41, No. 10, 1466-1477

Bose N.K. (1982) Applied Multidimensional Systems Theory, New York,. Van Nostrand Reinhold; (2000) private communication

Buchberger B. (1970) Ein algorithmishes kriterium fur die losbarkeit eines algebraishen gleichungsystems, *Aeq. Math.*, Vol. 4, 374-383

Caley Cambridge and Dublin Mathematical Journal, vol. III, p. 116

Chyi Hwang, Tong-Yi Guo, Leang-San Shieh (1993) A lattice realization for 2-D separable-denominator digital filters, *Circuits Systems Signal Processing*, Vol. 12, No. 3, 465-488

Cockburn J.C., Morton B.G. (1997) Linear fractional representations of uncertain systems, *Automatica*, Vol. 33, No.7, 1263-1271

Cockburn J.C. (1998) Linear fractional representations of uncertain systems using the structured-tree-decomposition approach (personal communication)

De La Sen M. (1997) On the hyperstability of a class of hybrid systems, *Int. J. Syst. Sci.*, Vol. 28, No. 9, 925-934

Doyle J. (1982) Analysis of feedback systems with structured uncertainties, *IEE Proc.-D*, Vol. 129, 242-250

Eising R. (1978) Realization and stabilization of 2-D systems *IEEE Trans. on Automatic Control*, AC-23, 793-799; (1979) Controllability and observability of 2-D systems, *IEEE Trans. on Automatic Control*, Vol. AC-24, No.1, 132-133; (1980) State-space realization and inversion of 2-D systems, *IEEE Trans. on Circuits and Systems*, Vol. CAS-27, 612-619

Fettweis A. (1984) Multidimensional circuits and systems In Proc. IEEE Int. Symp. on Circuits and Systems, Vol. 3 951-957; (1992) Simulation of hydromechanical partial differential equations by discrete passive dynamical systems In Kimura H. *et al* Eds.

Recent Advances in Mathematical Theory of Systems, Control and Signal Processing II, MITA Press, 489-494

Fettweis A., Nitsche G. (1991) Numerical integration of partial differential equations using principles of multidimensional wave digital filters, *Journal of VLSI Signal Processing*, vol. 3, 7-24 (1991).

Fornasini E. (1991) A 2-D systems approach to river pollution modelling, *Multi-dimensional Systems and Signal Processing*, Vol. 2, 233-265; (1999) private communication

Fornasini E., Marchesini G. (1978) Doubly-indexed dynamical systems, *Math. Syst.* Theory, Vol. 12, 59-72; (1991) A 2-D systems approach to river pollution modeling, *Multidimensional Systems and and Signal Processing*, Vol. 1, 233-265

Fornasini E., Zampieri S. (1991) A note on the state space realization of 2D FIR transfer functions, *Systems & Control Letters*, Vol. 16, 117-122

Gałkowski K. (1981a) The state-space realization of an n-dimensional transfer function, *Int. J. Circuit Theory and Applications*, Vol.9, 189-197; (1981b) Three-way companion matrices of three variable polynomials. *Linear Algebra and its Applications*. 37, 55-75; (1988) The similarity transformation of the Roesser model of N-D systems, Proc. of the IMA Conf. on Applications of Matrix Theory, Bradford, 1988, 193-202; (1991) An *a'priori* nonminimal state-space realization of n-D systems, *Linear Algebra and Its Applications*, Vol. 151, 185-198; (1992) Transformation of the transfer function variables of the singular n-dimensional Roesser model, *Int. J. Circuit Theory and Applications*, Vol. 20, 63-74; (1994) State-space Realizations of *n*-D Systems, Scientific Papers of the Institute of Telecommunications and Acoustics of the Technical University of Wrocław No. 76, Monographs No. 40; (1996a) Elementary operations and equivalence of 2-D systems, *Int. J.Control*, Vol. 63, No.6, 1129-1148; (1996b) The Fornasini-Marchesini and the Roesser model - Algebraic methods for recasting, *IEEE Trans. on Automatic Control*, Vol. AC-41, No.1; (1996c) Matrix description of multivariable polynomials, *Linear Algebra and Its Applications*, vol. 234, no.2, 209-226; (1997) State-space realizations of multi-input multi-output n-D systems - elementary operations approach, *Int. J. Control*, Vol. 66, No.1, 119-140; (2000a) State-space realizations of MIMO 2D discrete linear systems - Elementary operation and variable inversion approach, *International Journal of Control*, Vol. 73, No. 3, 242-253; (2000b) A perspective on singularity in 2D linear systems, *Multidimensional Systems and Signal Processing*, Vol. 11, No. 1-2, April, 83-108; (2001) Minimal state-space realization of the particular case of SISO nD discrete linear systems, *International Journal of Control*, submitted

Gałkowski K., Ed. (2000) Proceedings of 2-nd International Workshop on Multidimensional (nD) Systems (NDS-2000), Czocha Castle, Poland, June 27-30, 2000.

Gałkowski K., Piekarski M.S. (1978) The companion matrix for two variable polynomials, Ist Natl. Conf. Circuit Theory and Electronic Circuits, Podlesice 1977, Proceedings Vol.2, in Polish

Gałkowski K., Rogers E., Owens D, H. (1996) A new state-space model of linear discrete multipass process, *Bulletin of Polish Academy of Sciences, Technical Sciences*, Vol.44, No.1, 69-80; (1998) Matrix rank based conditions for reachability / controllability of discrete linear repetitive processes, *Linear Algebra and Its Applications*, vol. 275-276, 201-224

Gałkowski K., Rogers E., Gramacki A., Gramacki J., Owens D.H. (1999) Higher order discretization for a class of 2D continuous-discrete linear systems, *IEE Proceedings - Circuits, Devices and Systems*, Vol 146, No. 6, pp 315-320

Gałkowski K., Wood J., Eds. (2001) *Recent developments in nD systems*, Taylor & Francis, in print

Gatazzo R. (1991) Polynomial matrices with given determinant, *Linear Algebra and Applications*, Vol.144, 107-120

Gisladottir J.V., Lev-Ari H. and Kailath T. (1990) Orthogonal realization of first-order allpass filters for two-dimensional signals, *Multidimensional Systems and Signal Processing*, Vol. 1, 39-50

Goodman D. (1977) Some stability properties of two-dimensional linear shift-invariant filters, *IEEE Trans. on Circuits and Systems*, Vol. CAS 24, 201-208

Gregor J. (1993) Singular systems of partial difference equations, *Multidimensional Systems and Signal Processing*, Vol. 4, 67-82

Guiver J.P., Bose N.K. (1982), Polynomial matrix primitive factorization over arbitrary coefficient field and related results, *IEEE Trans. on Circuits and Systems*, Vol. CAS-29, No. 10, 649-657

Gundelfinger, Auflosung eines systems von gleihungen, worunter zwei quadratisch und die ubrigen linear, in *Schloemilch Zeitschrift fur Math. u Ph.*, Vol. XVIII p. 549

Guoxiang Gu, Aravena, Kemin Zhou (1991) On minimal realization of 2-D systems, *IEEE Trans. on Circuits and Systems*, Vol. CAS-38, No. 10, 1228-1233

Hinamoto T., Fairman F.W. (1984) Realizations of the Attasi state space model for 2D filters, *Int. J. Systems Sci.* Vol. 15, No. 2, 215-228

Johnson D.S., Rogers E., Pugh A.C., Hayton G.E., Owens D.H. (1996) A polynomial matrix theory for a certain class of 2-D linear systems, *Linear Algebra and its applications*, Vol. 241-243, 669-703

Jezek J. (1983) Conjugated and symmetric polynomial equations, II: Discrete-time systems, *Kybernetika*, Vol. 19, No. 3, 196-211; (1989) An algebraic approach to the synthesis of control for linear discrete meromrphic systems, *Kybernetika*, Vol. 25, No. 2, 73-82

Johnson D.S. (1993) Coprimeness in multidimensional system theory and symbolic computation, PhD thesis, Loughborough University of Technology, UK

Johnson D.S., Pugh A.C., Hayton G.E. (1992) On n-D matrix fraction descriptions, *Proc. American Control Conference-92*, Vol.1, 357-358

Johnson D.S., Rogers E., Pugh A.C., Hayton G.E., Owens D.H., 1996, A polynomial Matrix theory for a certain class of 2-D linear systems, *Linear Algebra and Its Applications*, no.241-243, pp 669-703.

Kaczorek T. (1982) New algorithms of solving 2-D polynomial equations, *Bull. Pol. Acad. Techn. Sci.*, Vol.30, No 5-6, 77-83; (1985) Two Dimensional Linear Systems, *Lecture Notes in Control and Information Sciences*, No. 68, Berlin: Springer-Verlag; (1987) Reduction of nD linear systems to 1D systems with variable structure, *Bulletin of Polish Academy of Science-Technical Sciences*, Vol. 35, No. 11-12, 623-631; (1988) The singular general model of 2-D systems and its solution, *IEEE Trans on Automatic Control*, Vol. AC-33, No.11, 1060-1061; (1989) Singular models of 2-D systems, Modeling and Simulation of Systems, P. Breedveld *et al* (editors) J.C.Baltzer AG, Scientific Publishing Co. IMACS, 389-392; (1990) Rational and polynomial solutions to 2-D polynomial matrix equations, *Bull. Pol. Acad. Techn. Sci.* , Vol.39, No 1, 105-109; (1991) Equivalence and similarity for singular 2-D linear systems. in New Trends in Systems Theory, Ed. by Conte G., Perdon A.M., Wyman B., Birkhauser, 448-455; (1992a) Linear Control Systems, Research Studies Press LTD, Taunton England, John Wiley & Sons, NY (1992b) Local controllability, reachability and reconstructibility of the general singular model of 2-D systems, *IEEE Trans on Automatic Control*, Vol. AC-37, No.10, 1527-1530; (1994) 2-D continuous-discrete linear systems In Proc. Tenth Int. Conf. on System Eng. ICSE'94, Vol. 1, 550-557; (1999) Equivalence of nD singular Roesser and Fornasini Marchesini models, *Bulletin of The Polish Academy of Sciences*, Vol. 47, No. 3, 235-246

Kailath T. (1980) Linear Systems, Prentice-Hall, Englewood Cliffs, N.Y.

Kawakami A. (1990) A realization method of two-dimensional rational transfer functions *IEEE Trans. on Circuits and Systems*, Vol. CAS-37, No. 3, 425-431

Klamka J. (1988) Controllability of dynamical systems, Kluwer, Dorttecht, The Netherlands

Kron G. (1965). Tensor Analysis of Networks. London: J. Wiley

Kummert A. (1989) Synthesis of two-dimensional loseless m-ports with prescribed scattering matrix, *Circuits Systems Signal Processing*, Vol. 8, No. 1, 97-119

Kung S.Y., Levy B.C., Morf M., Kailath T. (1977) New results in 2-D systems theory, part I (II), *Proc. IEEE*, Vol. 65, 861-872 (945-961)

Kurek J.E. (1985) The general state-space model for a two-dimensional digital systems, *IEEE Trans. on Automatic Control*, Vol. AC-30, 345-354; (1993) Observer realization for 2-D systems, *Bulletin of the Polish Academy of Sciences, Technical Sciences*, Vol. 41, No. 4, 381-390

Lancaster P. (1969) Theory of Matrices, Academic Press, N.Y.

Lewis F.L. (1992) A review of 2-D implicit systems, *Automatica*, Vol. 28, no.2, 345-354

Manikopoulos C.N., Antoniou G. (1990) State-space realizations of three-dimensional systems using the modified Cauer form, *Int. J. Systems Sci.*, Vol. 21, No. 12, 2637-2678

Mentzelopoulou S.H., Theodorou N.J. (1991) n-dimensional minimal state-space realization, *IEEE Trans. on Circuits and Systems*, Vol. CAS-38, No. 3, 340-343

Oldenburger R. (1940). Higher dimensional determinants. *Amer. Math. Monthly*. 47, 25-33.

Ortega J.M., Rheinboldt W.C. (2000) Iterative Solution of Nonlinear Equations in Several variables. SIAM, Philadelphia

Ozaki H., Kasami T. (1960) Positive real functions of several variables and their applications to variable networks, *IRE Trans. on Circuit Theory*, Vol. 7, 251-260

Paraskevopoulos, Kiritsis (1993) Minimal realisation of recursive ands nonrecursive three-dimensional systems, *IEE Proceedings-G*, Vol. 140, No. 3, 187-190

Park H., Woodburn C. (1995) An algorithmic proof of Suslin's stability theorem for polynomial rings, *Journal of Algebra*, Vol. 178, 277-298

Piekarski M. (1988) A state space synthesis of transfer matrix of orthogonal two-dimensional digital filters, Proc. of XI Natl. Conf. Circuit Theory and Electronic Devices, Łódź-Rytro 1988, 84-89, in Polish

Pontriagin L.S. (1955) On the zeros of some elementary transcendental functions, *Amer. Math. Soc. Transl.*, Ser. 2, Vol. 1, 95-110

Porter W.A., Aravena J.L. (1984) 1-D models for m-D processes, *IEEE Trans. on Circuits and Systems* Vol. 31, No. 742-744

Premaratne K., Jury E.I., Mansour M. (1990) Multivariable canonical forms for model reduction of 2-D discrete time systems, *IEEE Trans. on Circuits and Systems* Vol. 37, No. 4, 488-501

Pugh A.C., Johnson D.S. (1992) 2-D system structure and Applications, IFAC Workshop on System Structure and Control, Prague, 3-5 Sept.

Pugh A.C., Karampetakis N.P., Vardulakis A.I.G., Hayton G.E. (1994) A fundamental notion of equivalence for linear multivariable systems, *IEEE Trans on Automatic Control*, Vol. AC-39, (1994)

Pugh A.C., McInerney S.J., Boudellioua M.S., Hayton G.E. (1998) Matrix pencil of a general 2-D polynomial matrix, *Int. J. Control*, Vol. 71, No. 6, 1027-1050

Raghuramireddy D, Unbehauen R. (1993) Implementation of real coefficient two-dimensional denominator-separable digital filters based on decomposition techniques, IEE *IEE Proceedings-G*, Vol. 140, No. 1, 23-32

Roberts P.D. (2000) Numerical investigations of a stability theorem arising from the 2-dimensional analysis of an iterative optimal control algorithm, *Multidimensional Systems and Signal Processing*, Vol. 11, No. 1/2, 109-124

Rocha P. (1990) Structure and Representations for 2-D Systems, PhD Thesis, University of Groningen, The Netherlands

Rocha P. Rogers E., Owens D, H. (1996) Stability of discrete non-unit memory linear repetitive processes-a two-dimensional systems interpretation, *Int. J. Control*, Vol. 63, 457-482

Roesser R.P. (1975) A discrete state-space model for linear image processing, *IEEE Trans on Automatic Control*, Vol. AC-20, 1-10

Rogers E., Owens D.H. (1992) Stability analysis for linear repetitive processes, *Lecture Notes in Control and Information Sciences*, 175, Ed. Thoma M., Wyner W., Springer Verlag, Berlin

Rosenbrock H.H. (1970) State-Space and Multivariable Theory, London, Nelson

Sebek M. (1988) One more counter example in n-D systems – unimodular versus elementary operations, *IEEE Trans. Automatic Control*, Vol. AC-33, 502-503

Smyth K. (1992) Computer aided analysis for linear repetitive processes, PhD Thesis, University of Strathclyde, Glasgow, UK

Sokolov N.P. (1972). Introduction to the theory of N-way matrices (in Russian), Kiev. Naukova Dumka.

Sontag E. (1978) On first-order equations for multidimensional filters, *IEEE Trans. Acoust. Speech Signal Processing*, Vol. 26, No. 5, 480-482

Suslin A. A. (1976) Projective modules over a polynomial ring are free, *Soviet Math. Dokl.* Vol. 17, 1160-1164

Sylvester *Cambridge and Dublin Mathematical Journal*, Vol. VII, p. 68

Theodoru N., Tzafestas S.G. (1984a) A canonical state-space model for three-dimensional systems, *Int. J. Systems Sci.*, Vol. 15, No. 12, 1353-1379; (1984b) A canonical state-space model for *m*-dimensional discrete systems In Multivariable Control: New Concepts and Tools, S. Tzafestas Eds., D. Reidel, Dortrecht

Tzafestas S.G., Theodorou N. (1987) An inductive approach to the state space representation of multidimensional systems, *Control-Theory and Advanced Technology*, Vol. 3, No. 4, 293-322

Uetake Y. (1992) Realization of noncausal 2-D systems based on a descriptor model, *IEEE Trans on Automatic Control*, Vol. AC-37, 1838-1840

Uruski M., Piekarski M. (1972) Synthests of a network containing a cascade of commensurate transmissions lines and lumped elements, Proc. IEEE, Vol. 119, No. 2, 153-159

Xiao C., Sreeram V., Liu W.Q., Venetsanopoulos A.N. (1998) Identification and model reduction of 2-D systems via the extended impulse response Gramians, *Automatica*, Vol. 34, No. 1, 93-101

Varoufakis S.J., Antoniou G.E. (1991) Circuit and state-space realization of 2-D all-pass digital filters via the bilinear transform, *IEEE Trans on Circuits and Systems*, Vol. CAS-38, No. 9, 1104-1107

Venkateswarlu T., Eswaran C. and Antoniou A. (1990) Realization on multidimensional digital transfer functions, *Multidimensional Systems and Signal Processing*, Vol. 1, 179-198

Wood J., Rogers E., Owens D. (1998) A formal theory of matrix primeness, *Mathematics of Control, Signals, and Systems*, Vol. 11, 40-78

Youla D.C., Gnavi G. (1979) Notes on n dimensional systems, *IEEE Trans on Circuits and Systems*, Vol. CAS-26, No. 2, 105-111

Youla D.C., Pickel P.F. (1984) The Quillen-Suslin theorem and the structure of n-dimensional elementary polynomial matrices, *IEEE Trans. on Circuits and Systems* Vol. CAS-31(6):513-518

Zieliński J.S. (1990). Power system state identification with the three-dimensional (3-D) matrices. *Archive of Electrotechnics*, XL, 1/4, 375-392

Żak S. H., Lee E. B., Lu W. S. (1986) Realizations of 2-D filters and time delay systems, *IEEE Trans on Circuits and Systems*, Vol. CAS-33, No. 12, 1241-1244

Index

Lecture Notes in Control and Information Sciences

Edited by M. Thoma and M. Morari

1997–2000 Published Titles:

Vol. 240: Lin, Z.
Low Gain Feedback
376 pp. 1999 [1-85233-081-3]

Vol. 241: Yamamoto, Y.; Hara S.
Learning, Control and Hybrid Systems
472 pp. 1999 [1-85233-076-7]

Vol. 242: Conte, G.; Moog, C.H.; Perdon A.M.
Nonlinear Control Systems
192 pp. 1999 [1-85233-151-8]

Vol. 243: Tzafestas, S.G.; Schmidt, G. (Eds)
Progress in Systems and Robot Analysis and
Control Design
624 pp. 1999 [1-85233-123-2]

Vol. 244: Nijmeijer, H.; Fossen, T.I. (Eds)
New Directions in Nonlinear Observer Design
552pp: 1999 [1-85233-134-8]

Vol. 245: Garulli, A.; Tesi, A.; Vicino, A. (Eds)
Robustness in Identification and Control
448pp: 1999 [1-85233-179-8]

Vol. 246: Aeyels, D.; Lamnabhi-Lagarrigue, F.;
van der Schaft, A. (Eds)
Stability and Stabilization of Nonlinear
Systems
408pp: 1999 [1-85233-638-2]

Vol. 247: Young, K.D.; Özgüner, Ü. (Eds)
Variable Structure Systems, Sliding Mode
and Nonlinear Control
400pp: 1999 [1-85233-197-6]

Vol. 248: Chen, Y.; Wen C.
Iterative Learning Control
216pp: 1999 [1-85233-190-9]

Vol. 249: Cooperman, G.; Jessen, E.;
Michler, G. (Eds)
Workshop on Wide Area Networks and High
Performance Computing
352pp: 1999 [1-85233-642-0]

Vol. 250: Corke, P. ; Trevelyan, J. (Eds)
Experimental Robotics VI
552pp: 2000 [1-85233-210-7]

Vol. 251: van der Schaft, A. ; Schumacher, J.
An Introduction to Hybrid Dynamical Systems
192pp: 2000 [1-85233-233-6]

Vol. 252: Salapaka, M.V.; Dahleh, M.
Multiple Objective Control Synthesis
192pp. 2000 [1-85233-256-5]

Vol. 253: Elzer, P.F.; Kluwe, R.H.;
Boussoffara, B.
Human Error and System Design and
Management
240pp. 2000 [1-85233-234-4]

Vol. 254: Hammer, B.
Learning with Recurrent Neural Networks
160pp. 2000 [1-85233-343-X]

Vol. 255: Leonessa, A.; Haddad, W.H.;
Chellaboina V.
Hierarchical Nonlinear Switching Control
Design with Applications to Propulsion
Systems
152pp. 2000 [1-85233-335-9]

Vol. 256: Zerz, E.
Topics in Multidimensional Linear Systems
Theory
176pp. 2000 [1-85233-336-7]

Vol. 257: Moallem, M.; Patel, R.V.;
Khorasani, K.
Flexible-link Robot Manipulators
176pp. 2001 [1-85233-333-2]

Vol. 258: Isidori, A.; Lamnabhi-Lagarrigue, F.;
Respondek, W. (Eds)
Nonlinear Control in the Year 2000
Volume 1
616pp. 2001 [1-85233-363-4]

Vol. 259: Isidori, A.; Lamnabhi-Lagarrigue, F.;
Respondek, W. (Eds)
Nonlinear Control in the Year 2000
Volume 2
640pp. 2001 [1-85233-364-2]

Vol. 260: Kugi, A.
Non-linear Control Based on Physical Models
192pp. 2001 [1-85233-329-4]

Vol. 261: Talebi, H.A.; Patel, R.V.;
Khorasani, K.
Control of Flexible-link Manipulators Using
Neural Networks
168pp. 2001 [1-85233-409-6]